Unfinished Animal

By Theodore Roszak

Unfinished Animal

THE AQUARIAN FRONTIER
AND THE EVOLUTION
OF CONSCIOUSNESS

Theodore Roszak

Harper & Row, Publishers

NEW YORK
EVANSTON
SAN FRANCISCO
LONDON

1817

FIRST EDITION

Designed by Sidney Feinberg

Library of Congress Cataloging in Publication Data

Roszak, Theodore, 1933–
  Unfinished animal.
  Includes bibliographical references and index.
  1. Religion. 2. Evolution. I. Title.
BL48.R56 1975   200   75–9333
ISBN 0–06–067016–9

---

75 76 77 78 79 10 9 8 7 6 5 4 3 2 1

Sail forth—steer for the deep waters only,
Reckless O soul, exploring, I with thee, and thou with me,
For we are bound where mariner has not yet dared to go,
And we will risk the ship, ourselves and all.

—Walt Whitman, *Passage to India*

The bomb has already dropped, and we are the mutants.
—Berkeley, California graffito

For Kathryn
who has known since our days at Harry Hall
that music and make-believe tell all the
deepest truths

Contents

Unfinished Animal

INTRODUCTION: Pico's Chameleon and the Consciousness Circuit

> Every one my age that I know who has no decent job just feels like we're waiting for something—some series of events—like it's not quite time to start our lives or get into too much, cause we'll have our hands full in a little while. Can you feel it? Does that come across to you? We're only here to put in time till . . .

"till . . ." And then the thought melts away. Into what? A vacuum of yearning? A hope too desperate to be uttered?

The words are from a letter written by a former student of mine, a young woman who now survives meagerly on food stamps in a California coastal retreat while devouring an exclusive reading diet of science fiction and pop occultism.

Revolutionary spirits of my father's generation waited for Lefty. Existentialist heroes of my youth waited for Godot. Neither showed up. Still, for people like my misty-minded young friend, the waiting game continues, though its mood has much changed—from anxious to wistfully apocalyptic. Asked what she is waiting for, I suspect from other conversations that she would say she is waiting for Atlantis to rise from the sea, for the UFOs to descend, for the star gates to open, for a new avatar to sweep in from the mystic East trailing clouds of ecstasy across the land.

And meanwhile . . . one brews Mu tea, attends to the rhythm of one's cakras, and travels the "consciousness circuit." From Zen to Tantra, from LSD to astral projection, from the primal scream to the Divine Light Mission, reporting back wonders and astonishments all along the way.

1

Would she believe what she reports? Do any of them believe what they have to tell us about their experiences of the extraordinary, the growing number who have withdrawn to the countercultural fringes of our society to confabulate with Sabian signs and aura readings, Tarot cards and yarrow stalks?

Yes and no. One can only be ambiguous. The occult and mystic (often liberally laced with science fiction and comic book fantasy) have become special vehicles of popular culture in our day, a coded vocabulary suspended between caprice and conviction, between a teasing playfulness and a deadly earnest. What they mean to say is not always what they literally say. Often enough, they are a private rebellion against conventional intellect. Follies intended to confound the wise. But that is not all they represent. They are also, and more importantly, premonitions of the miraculous, an instinctive sense of extraordinary new realities hovering above our time like the flames of Penetecost, demanding glossalalic celebrations incomprehensible to the uninitiated.

Within the past few years, I have found myself more and more in the company of people like my former student: bright, widely read, well-educated people whose style it has become to endorse and accept all things occultly marvelous. In such circles, skepticism is a dead language, intellectual caution an outdated fashion. That Edgar Cayce could diagnose the illnesses of distant patients and predict earthquakes by psychic readings . . . that the pyramids were built by ancient astronauts . . . that orgone boxes can trap the life energy of the universe . . . that the continents of the Earth were settled by migrations from lost Lemuria . . . one does not question these reports, but, calmly letting the boundary between fact and fairy tale blur, one *uses* them—uses them to stretch one's powers of amazement. One listens *through* them to hear still another intimation of astounding possibilities, a shared conviction which allows one to say,

"Yes, you feel it too, don't you? That we are at the turning point, the *kairos*, where the orders of reality shift and the impossible happens as naturally as the changing of seasons."

Such intellectual permissiveness risks a multitude of sins, not the least of which is plain gullibility. Certainly in assessing this venturesome and compliant habit of mind, we must remain dutifully alert to the puerile follies that collect at its extreme; in the pages that follow, I will have some hard words to spend on its many banalities. Yet, with that caution duly noted, I remain persuaded that we deal here with the popular reflection of a significant new sensibility, one whose higher implications have already begun to emerge in more mature heads. At least to so much this book commits itself:

1. that we find in this rising curiosity for the marvelous, the popular unfolding of an authentically spiritual quest, which has been driven into a variety of unorthodox channels by the rigidity of conventional religion in the Western world. While that quest has yet to develop the discriminating taste it will require if it is to endure and flourish, its emergence is the necessary and healthy sign of spiritual resurgence.

2. that we can discern, through all these starry-eyed images of an Aquarian Age filled with wonders and well-being, a transformation of human personality in progress which is of *evolutionary* proportions, a shift of consciousness fully as epoch-making as the appearance of speech or of the tool-making talents in our cultural repertory.

What follows, then, is a survey and critique of the current religious revival in Western society, particularly with respect to its ethical and political implications, together with an exploration of its meaning as a stage in our evolutionary growth.

It is with the second of these subjects that we will spend the most time, because, along the consciousness circuit, evolutionary transformation is rapidly becoming the predominant interpretation of our historical moment. Among a growing number of those who move with the forward currents of psychotherapy and the healing arts, "consciousness research" and the new religions, a spontaneous consensus has sprung up around the evolutionary image of human potentiality. It has very nearly become a popular mythology of the day, this sense that we stand in witness to a planet-wide mutation of mind which promises to liberate energies of will and resources of vision long maturing in the depths of our identity. The new ecological awareness, together with its sense of allegiance to the planet as a whole, is one sign of the great transition. The rapid convergence of age-old spiritual disciplines and contemporary psychotherapy is another. Both follow with dizzying speed upon the advent of a world-revolutionary politics which aspires to extend, at a stroke, the full dignities of personhood across the entire human spectrum, to all who have been excluded and despised, every race, class, sex, culture, lifestyle. Suddenly, as we grow more introspectively inquisitive about the deep powers of the personality, our ethical concern becomes more universal than ever before; it strives to embrace the natural beauties and all sentient beings, each in her and his and its native peculiarity. Introspection and universality: center and circumference. Personal awareness burrows deeper into itself; our sense of belonging reaches out further. It all happens at once, the concentration of mind, the expansion of loyalty. As if consciousness, probing the black hole of the mind, had come out at the all-encompassing perimeters of nature. Only the fantasies of surrealist imagination in our art, song, and literature seem able to embrace the dimensions of such a cultural transition.

But if I choose the evolutionary image here as the key to understanding our bewildering times, this is not to endorse every casual and careless use that is made of the idea. It is rather to deepen the concept by connecting it once more with the long forgotten spiritual traditions from which it derives. And to do this is to suggest a meaning for the evolution of consciousness which cannot be contained by biology or neurophysiology or any form of psychotherapy, a meaning which many of the carriers of this evolutionary destiny may not yet have recognized. My argument will be that the popular mythology of evolving consciousness, now so awkwardly overlaid with scientific and psychotherapeutic trappings, echoes another, far older myth, a story of redemption and transcendence which carries us back to the dawn of religious awareness. Take the myth at its full value, and it reminds us of the task that has been laid upon us as nature's uniquely self-creating, self-defining species: to discover the godlikeness in whose image we are said to have been cast. After some fashion that I find myself unable to confine within the boundaries of myth or history, scientific fact or metaphysical metaphor, that task now enters our remembrance as an urgent vocation alongside all the dire social and political necessities of the day: their companion challenge and, I think, their indispensable complement.

Lewis Mumford tells us that in every historical era there are dominant themes and emergent themes. In our time, the dominant themes are science, secular humanism, global industrialism, and social revolution. But the emergent theme of the age has been sounded, I believe, by those who begin to see themselves as unfinished animals summoned to unfold astonishing possibilities. For all the seeming irrelevance and perversity of their style, those who step to the Aquarian tempo of life may sense with a unique vividness that our much endangered interval in history demands more of us than mere survival (as much as that might be to ask), more than social revolution (as necessary as that might be). It demands a regeneration of life at some finer, more vibrant level of being, a qualitative great leap forward of the species whose outcome we can only fantastically prefigure by outlandish assertions of the strange and awesome.

The powers of the third eye . . . the secret of levitation . . . the advent of extraterrestrial wizards . . . all these are rudimentary images of collective self-transcendence, gathered haphazardly from sources high and low, wise and wild. We who have been waiting so long for the bomb to fall—or not to fall—what choice have we but to set our sights upon a salvation as apocalyptic as the doom that hounds us?

But let us step back for a moment from the busy congestion and rapid turnover of the consciousness circuit and give our subject matter a little historical breathing space. As they appear in the pages of so many modish periodicals and trendy best sellers, or as the "therapy-and-awareness" smorgasbord of the local Growth Center and university extension, the psychospiritual fascinations of the day can easily be mistaken for an ephemeral amusement of the affluent society: a sort of mystic chic. But if we take the consciousness circuit, at its most significant, to be an effort to free the transcendent powers of the personality from the dead hand of the culture's secular and religious orthodoxies, then Western society—or at least one importantly dissenting fringe of it—has been negotiating a crucial phase of that circuit for the past five centuries, on and off. To take a conveniently conspicuous point of departure, say ever since Pico della Mirandola issued his *Oration on the Dignity of Man* just five years before Columbus set sail for the New World.

In that famous declaration, the young Pico (only twenty-four years old at the time) described the human being as a creature of indeterminate and infinitely plastic nature: a "chameleon" capable of imitating all living forms high and low, angelic and demonic. He imagines God telling Adam:

Neither a fixed abode, nor a form that is thine alone, nor any function peculiar to thyself have we given thee, Adam, to the end that according to thy longing and according to thy judgement thou mayest have and possess what abode, what form, and what functions thou thyself shalt desire. . . . We have made thee neither of heaven nor of earth, neither mortal nor immortal, so that with freedom of choice and with honor, as though the maker and molder of thyself, thou mayest fashion thyself in whatever shape thou shalt prefer. Thou shalt have power to degenerate into the lower forms of life which are brutish. Thou shalt have the power, out of thy soul's judgement, to be reborn into the highest forms, which are divine.

And Pico finishes with the invitation:

Let a certain holy ambition invade our souls, so that, not content with the mediocre, we shall pant after the highest, and—since we may if we wish—toil with all our strength to obtain it. Let us disdain earthly things, despise heavenly things, and finally esteeming less whatever is of the world, hasten to that which is beyond the world and nearest to the Godhead.[1]

In Pico's statement, we have, for the first time in the modern West, a vision of human nature as unfulfilled potentiality, of life as an adventure in self-development. Humanness, Pico tells us, is not a closed box, but an open door . . . leading to an open door . . . leading to an open door. And he invites us to make our way through all these doors, discriminately experiencing the fullness of our identity. For Pico, the human condition is not what it has always been for both conventional Christianity and conventional humanism: a sternly prescribed choice between two airtight compartments, one the province of absolute virtue, the other the province of absolute evil: nature against grace, reason against the irrational, sanity against madness. Rather, he asks us to see ourselves as a grand spectrum of possibilities

1. Pico della Mirandola, *Oration on the Dignity of Man,* trans. Elizabeth Livermore Forbes (Chicago: University of Chicago Press, 1948).

whose unexplored regions include the godlike as well as the diabolical.

Had Pico's program for human development become, as he wished, the educational standard of our culture, Western society might have freed itself from the literalism and dogmatic intolerance of Christian orthodoxy, without rushing into the dismal materialism that dominates our scientific world view. We might have found our way to a new culture of the spirit, open to universal instruction, grounded in experience, capable of liberating the visionary dimensions of the mind. But the fate of Pico's way was to become a dissenting countercurrent to the cultural mainstream: either a saving remnant or a lunatic fringe, depending upon one's viewpoint.

The Renaissance stands as the first of three cultural eras in the modern West in which we find prominent creative minorities working along that countercurrent to expand the limits of the personality, with nothing less than godlikeness as their goal. The second was the Romantic movement of the late eighteenth to early nineteenth centuries. The third (and I suspect climactic) episode unfolds now in our own generation. For all the differences between these three periods, they are strikingly similar in the lifestyle and psychological tone of their most innovative figures: supremely anxious eras, yet supremely ambitious and energetic; painfully dislocated, yet relishing the freedom that comes of dislocation; torn with self-doubt, yet fired by a flamboyant taste for innovation and discovery. Each of the three celebrates personal originality and the eccentricities of genius, placing a special and urgent value on the inspired individual as culture maker. Each runs to extremes of discontent, nihilism, and demonic imbalance. Yet each has vastly enlarged our knowledge of human capacities, and in no respect more significantly than by way of its fascination with occult studies and exotic religion.

Conventional history remembers the Renaissance for its

rediscovery of classical antiquity. But it was also, and with no less enthusiasm, the era that rediscovered the subterranean veins of magic and mysticism in our culture. Its major intellectual contributions include the work of Marsilio Ficino's Florentine Academy, where Neoplatonism, Hermeticism, and the Christian Kabbala were featured studies. The Renaissance reclaimed numerology, astrology, and the Tarot from their heretical reputation and made alchemy both the handmaiden and the rival of early Western science. Pico himself was an avid student of astrology, Christian Kabbalism, and Neoplatonic philosophy. Indeed, nearly all the major intellectual figures of the period—including many of the founding fathers of modern science down to Newton's day—were deeply invested in such esoteric traditions, as if they believed there might lay hidden in these buried sources secrets of human nature and the universe that were nowhere else to be found. We could have no better image to personify the age than Marlowe's Doctor Faustus, pursuing the glamor of forbidden knowledge wherever it might lead, crying, " 'Tis magic, magic that hath ravished me" . . . all the way to the gates of perdition.

The same insatiable appetite for the esoteric, strange, and mysterious reemerges in the Romantic era as a rebellion against the cautious self-control of the Enlightenment. Once again, the Faustian image appears as the emblem of the age, this time in the shape of Goethe's demon-driven hero hungering for inncessant experience, for those heights and depths of knowledge which God and the devil guard as their own. The Romantic artists especially carried the search for buried potentialities to new lengths. They not only reached out to embrace the culture of exotic times and places, they also ransacked their own lives for extraordinary states of experience. Dream and nightmare, madness, childhood fancy, mania, frenzy, supernatural possession, suicidal depression, narcotic trance, visionary inspiration . . . it is thanks to the

Romantics, and especially to those who followed Baudelaire in regarding the goal of art as the maximum derangement of the sensibilities, that these outcast varieties of experience were admitted to the repertory of Western art and thought— though they also brought with them pathological and decadent elements that have done more to inform psychiatry than to foster creativity.

Most important of all, the Romantics recognized the uncanny as a definitive feature of nature—a cosmic reality that transcends the limits of mere abnormal psychology. And this soon brought them, like the Hermeticists and Christian Kabbalists of Pico's day, into the realm of occult and mystic studies. It is in the wake of the Romantic movement that sects and societies devoted to esoteric tradition reappear in Western society to resume the age-old quest for the philosopher's stone, the lost Shekinah, the alien God. Eliphas Lévi's circle in Paris, Madame Blavatsky's Theosophical Society, Gerard Encausse's L' Ordre des Silencieux Inconnus, Rudolf Steiner's Anthroposophy, A. E. Waite's Order of the Golden Dawn, Aleister Crowley's Order of the Silver Star . . . without endorsing all that transpired in these early occult groups, one can see them as significant conduits leading into the consciousness circuit.

Now, in our own generation, we are once again challenging the further reaches of the mind, but this time the project advances on the broadest possible front and with more knowledgeable borrowing (especially from the Oriental and primitive cultures) than ever before. One need only survey any well-stocked bookshop to appreciate the spectacular variety of exotic experience that has flooded the cultural mainstream in recent days—not only in the form of respectable scholarship, but even, or perhaps preferably, in reports of the most apocryphal, bizarre, and fabulous character. All the mystic odysseys and occult rhapsodies of the ages are suddenly before us rescued from the outer darkness of long neglect, together with countless yogas and therapies of radical introspection, astral excursions, manuals of meditation and do-it-yourself ecstasies, sorcerer's cookbooks, secrets of the holy temple, explorations of the diabolical and forbidden. Nobody who ever talked with a witch or walked with a zombie goes without an attentive audience these days; the market for supernatural amazements is inexhaustible and infinitely indulgent.

Nor is it mere academic curiosity that drives the quest forward, but, as with the Romantics, a passion to experiment and participate—to gain immediate, personal access to the *terra incognita* of the human mind, whether by way of psychotherapy, psychedelic drugs, laboratory research, or authentic spiritual discipline. Never before has any society assembled so many materials from which to fashion its portrait of the human being. We are, as Robert Duncan has put it, congregating a "symposium of the whole" where nothing human shall be treated as outcast or unworthy, where all must be experienced and somehow made sense of.

I cannot speak for other cultures and their development; but for our own, I believe this protracted period of psychic exploration and experiment has been Western society's troubled passage through a crucial stage in the evolution of the human race. Having traveled further in the direction of spiritual alienation than any other people in history (and having drawn the rest of the world after us in our course), it now becomes our task to reclaim the human potentialities we have denigrated and sacrificed along the way. Through a thousand follies, false starts, and disasters, Pico's chameleon has been feeling its way toward a higher and finer identity, a transcendent condition of being which has been, since time out of mind, the supreme object of religious aspiration. Across the last five hundred years of our history, many of the brightest minds of our society have been ripening toward that condition, borrowing insights, images, and disciplines of enlightenment from every available source, courting the charges of heresy and madness, risking the demonic to draw nearer the godlike. In the absence of a robust and trustworthy spiritual authority in the modern West, the task has fallen mainly upon the gifted (often eccentrically gifted) few, figures like Pico and Boehme, Blake and Baudelaire, Jung and Joyce, Gurdjieff and Blavatsky, Buber and Yeats, Rilke and Steiner, Emerson and Whitman—mavericks who pitted themselves against both the diminished spirituality of conventional Christianity and the advance of secular humanism.

Now, within the last decade, the fragile cause of these oddballs and outsiders has taken on the dimensions of a major cultural movement, a hasty psychic migration into every unexplored corner of the personality. Very likely it is the biggest introspective binge any society in history has undergone. Of course, it is all happening too rapidly, too indiscriminately, and on too vast a scale. Every spiritual tradition in the human repertory is suddenly being asked to democratize its mysteries. And where that does not happen rapidly enough, we get dazzling personal improvisations or facile contemporary substitutes. But then perhaps this is how major evolutionary transitions happen. As the occasion grows ripe, there is an impetuous outburst of experiments and wild approximations, disastrous wrong-turnings and outright monstrosities. Until, out of the welter, there emerges a rich mix of possibilities, the scrambled elements of a new breed awaiting selection and synthesis within the controlling pattern of development.

This book hardly undertakes to prescribe or predict the synthesis our predicament requires; but it does seek to identify some of the monstrosities on the scene and to suggest a few principles of selection. We have clearly reached a stage in our psycho-spiritual explorations where discriminating

among the countless options before us is the imperative
need. We may stand on the threshold of a grand transformation, but there is no guarantee we will make our way across.
For we may, after all, be a failed species, doomed, despite
the example and instruction of the great masters who have
worked among us, to bungle the opportunity at hand. If we
do indeed confront the choice before us as free spirits in the
universe, then we are as free to lose as to win.

It is *spiritual intelligence* the moment demands of us: the
power to tell the greater from the lesser reality, the sacred
paradigm from its copies and secular counterfeits. Spiritual
intelligence—without it, the consciousness circuit will surely
become a lethal swamp of paranormal entertainments, facile
therapeutic tricks, authoritarian guru trips, demonic
subversions.

But where is spiritual intelligence to be found, especially
in *this* society whose peculiar history renders it as incompetent at dealing with the subtleties of the spiritual life
as the Bushman-Hottentots would be at programing a computer? The answer that suggests itself at once to my own
tastes is: we must find it in sacred tradition, in those ancient
springs of visionary knowledge which are the source of the
mystic and occult schools, and from which we draw our entire repertory of transcendent symbolism and metaphysical
insight. The "perennial wisdom," the "secret doctrine," the
"old gnosis" . . . if the idea of such an original and universal
epiphany is a "myth," then it is one of the *good* myths; in
fact, the myth which underlies our very conception of truth
as that to which all people voluntarily acquiesce in their common humanity. To the degree that the consciousness circuit
has encouraged a new respect for religious tradition, I believe it has made an invaluable contribution to our enlightenment, not only in winning back from obscurity a
wealth of spiritual materials, but in blunting the cultural

egotism of those who insist upon regarding all "paranormal" experience as a sweet research project for laboratory psychology. As it is, the scene is worrisomely crowded with encephalographs; we need more than a gentle reminder that our scientized habits of thought may be more the problem than the solution.

Yet, faced with the terrible confusions that mark the freewheeling religious revival we are in, those who possess the gift of spiritual intelligence may well see fit to respond by retreating to a well-fortified elitism. Contempt, austere withdrawal, cold unapproachability, these are also temptations of the moment, as much as sensational debasement. I work from the assumption that there remains with us, even in this era of fanatic secularization, a scattering of enlightened souls among whom sacred teachings and the visionary energies live on in relative purity. I refer not only to the oriental masters and sanghas now being so eagerly sought out, but to gifted minds and surviving spiritual communities in our own society: metaphysicians, contemplatives, savants who have weathered the skeptical blight of the modern period and have much to offer in the way of recycling our religious heritage. Should they retreat from us into a forbidding isolation, our loss (and I think theirs) would be irreparable. Undeniably, such a protective strategy enjoys the sanction of tradition; but the crisis we are now in is wholly unanticipated by tradition. There is a finality about our choices which only the wilfully blind or hopelessly haughty could fail to see. We—all of us who set out on this strange evolutionary odyssey—need reliable instruction with an unprecedented urgency; and in so far as those who guard the traditions are duty-bound to preserve and give over their inheritance, *they need us* with exactly the same urgency. For when do they expect so pregnant an opportunity to come again? Indeed, if the present occasion is lost, when afterwards do they expect to find any audience at all?

These days, people begin their quest along many bizarre roads. They grope, they stumble often, they follow false lights, they too easily mistake the theater for the temple, the circus for the sanctuary. Members of a culture which has seen many of its finest minds insist upon militant godlessness as the only morally responsible stance in the world, they may not always know or care to admit what destination they really long to reach. But people need not finish where they begin. That so many have so suddenly begun at all must give us reason for hope. Yes, we must preserve our sense of discrimination; but we owe ourselves and our fellows much mercy, and no little patience. There are some I have met—men and women deeply studied in esoteric traditions—who flinch back into guarded reticence if they once sense their audience has grown beyond eight close friends. They have grown too fond, I think, of the closet and the cloister, perhaps forgetting that there is a fastidiousness of personal taste which is as great a betrayal of the truth as tasteless vulgarization. In my own education, no writer has been more helpful as an example of keen, spiritual discrimination than the French Sufi, René Guénon. But here was a mind whose very precision led to an aristocratic intolerance and an elitism that risked sterility. (One almost suspects that, were he alive today, Guénon would be horrified to find his work available to the millions in the "new consciousness" sections of paperback bookstores everywhere.) On the other hand, I think of Thomas Merton as the example of one whose commitment to tradition (and to the strictest form of Catholic monasticism at that) never prevented him from belonging compassionately to his times as concerned critic and counselor. Even in hermitical retreat at his Trappist monastery, he remained an engaged man in constant and kindly dialogue with Marxist and mystic, beatnik and Buddhist. Where would we be without his kind?

It is a dilemma that has always confronted those who

watch over the mysteries: where does one fix just the right public perimeter for those who approach, either in curiosity, reverence, or doubt? Given my own limitations and many false starts in life, I can only hope that the good gurus among us will give that perimeter a generous size, recognizing in our aspirations at least a raw material they can work with.

While, as I hope this book makes clear, I am far from agreeing that "anything goes," my sympathies obviously belong to those who are trying to build bridges between our flawed, contemporary gestures of transcendence and the primordial sources of spiritual growth: teachers, artists, therapists, scholars, skillful in means, who are willing to let tradition borrow a modern vocabulary, while never forgetting how often we must return to the original to recapture what has been lost in the translation. Indeed, the evolutionary theory of visionary insight I present in chapter six requires that our collective, cultural experience be seen as just such a conversation between time and eternity: an uncanny fabric in which the profane world of our daily life is woven from a multitude of fine sacred threads. Drawn as I am by the view that transcendent possibilities lie at the root of every human fascination which is not irredeemably demonic, I think our accent must be upon experiment, outreach, and the patient cultivation of what may be history's last harvest of the spirit. What we need is a way to make sacred tradition our own, a music that adapts the old theme to the rhythm of our peculiar cultural style and experience. Our conviction must be that if we attend to our spiritual needs with sufficient depth and with absolute diligence, avoiding desperate decisions and sensational distractions, the old gnosis will find its own way to speak to us. We must believe that there exists, at the core of the mind, a shaft of subjectivity so deep that it can lead us clean through all that is perishably personal and culturally contingent to a universal ground beneath.

For twentieth-century Western society, such a descent into the deep mind at once suggests recourse to psychiatry. And it may well be that psychiatry—or psychotherapy in its broadest experimental sense—will prove the most promising way for many modern Westerners to approach the traditional teachings. That, at least, is the assumption this study makes in borrowing from psychotherapy its key images: that salvation is an inward quest of the individual, that the road to the visionary realities begins in the deep consciousness. Certainly we find in therapy the personalism and the mercy which can alone minister to our spiritual anguish, together with a fidelity to uncensored experience which our mainstream religions lack. And, too, we find in the wisest therapies the sense of disciplined commitment.

But what therapy must yet learn from the spiritual traditions, whose techniques it borrows ever more freely, is that the inner universe it studies is not *merely* psychological. It must recognize that the personal can serve as a point of entry into the transpersonal. And where that happens, our measure of normality must take on the new dimension of transcendence, becoming a higher sanity than psychiatry has yet contemplated.

We are only beginning to create therapies of the visionary mind in our society, strange marriages of modern Western psychology and ancient spiritual insight. Progressively, Western psychiatry is becoming the junior partner in these alliances. But, at the same time, the new therapies find it possible to deal eclectically among the often ethnospecific religious traditions, drawing from each its universal elements and so working toward a synthesis of insights and disciplines that befits our status as a planetary culture.

I suspect that for some time to come, until an all-transforming synthesis is achieved, spiritual intelligence in our society will stem from imaginative fusions of psychotherapy and traditional wisdom. Therapy will universalize the traditions; the traditions will deepen therapy and provide its metaphysical foundation. And perhaps, with the aid of both, this generation of chameleons will find the enlightenment it needs to complete its evolutionary adventure.

1

Aquarian Frontier

Religion by Any Other Name

In the eclipse of God, we have no place to begin but with ourselves. *Within* ourselves. All we have lost in the course of becoming this torn and tormented creature called modern man—the visionary energies, the discipline of the sacred—we discover again in the depths of our identity. There or not at all. We tunnel our way home through the floor of the mind, through that uncharted Self which is our other and inward universe.

In other times and in other cultures, it would go without question that to venture the further reaches of human identity is a religious exercise. As St. Augustine discovered, the depth dimension of autobiography is a prayer of divine communion: "Thou madest us for Thyself and our heart is restless till it rest in Thee." But, in the modern West, religion has long since become too crabbed and cautious a category to hold the burgeoning varieties of psychic exploration we see about us. For too many sensitive minds, "religion" means the censorship of experience, not its liberation. The very word groans with mind-binding orthodoxies. So they turn to exotic cultures, primitive or oriental, extracting from their traditions of worship the introspective techniques and disciplines of self-realization which our society has so rarely associated with religion that we have had to improvise a secular science—psychiatry—to provide for the need. Or the spiritual impulse of the time steps beyond the boundaries of religious tradition all together to become psychotherapy, sensory awareness, parapsychology, consciousness research, psychedelic tripping, bioenergetics, occult science. We are in the midst of a religious renaissance, but it is religion by any other name and every other name that binds our attention.

Look carefully enough and you are apt to discover an essentially religious quest under way in the most improbable pursuits on the contemporary scene. Unconventional medicine, for example. While faith cures and miraculous healing have always been counted among the hallowed powers, when have we ever had so many fully elaborated systems of occult medicine under intensive, popular study in our society, and when have the mysteries of the organism ever been so all-involving and participative a preoccupation? Reflexology, zone therapy, radionics, biotonics, autogenesis, herbology, biorhythms, aura readings, and a dozen kinds of massage—such methods of etherealized healing go well beyond the old camp meeting laying-on-of-hands. There is introspective involvement and metaphysical structure to their study. In fact, they rapidly become quasi-spiritual disciplines that employ the body as their sacramental object.

It is precisely this mystic elaboration of the organism that gains a prominent healer like Jack Schwarz his following. Schwarz, originally from Holland and now working out of Selma, Oregon, currently travels a lecture-demonstration circuit that takes him through several leading American medical schools. On the surface, his experiments in the "voluntary control of internal states" (pain, bleeding, infection) look like conventional research in psychosomatic medicine.

But enroll in his classes, and you soon find yourself caught up in an eclectic mixture of yoga, Tantric visualizations, and occult cosmologies. Draw closer to the man, and he takes on the aura of shaman, sage, and spiritual counselor; clinical research and mere medical science are soon left far behind; the discussion of physiology and disease broadens rapidly into visionary metaphysics and personal philosophy.

The same air of hygienic religiosity pervades the new, mystic athletics, like that practiced at the Esalen Institute Sports Center in San Francisco. Initiated in 1972 as a forum for discussing the vices of professionalism in sport, the Center quickly became a gathering place for those who have come to regard athletics as a contemplative therapy of the body and soul—in the spirit of aikido or t'ai chi ch'uan. It is difficult to imagine anything that more confounds the traditional Western notions of sacred and profane than such a mingling of physical culture and spiritual growth; yet, in his book *Golf in the Kingdom*, Michael Murphy, one of the founders of the Center, nicely adapts the mystic intention of these oriental exercise meditations to the game of golf. And, perhaps under the influence of Mike Spino's classes at Esalen in "Running as a Spiritual Experience," there are joggers I know who have turned the health fad of aerobics into a breath meditation modeled after pranayama yoga. For George Leonard, who has attempted the most ambitious philosophical elaboration of spiritualized sport and exercise, the "hidden dimension" of athletics "involves entering the realm of music and poetry, of the turning of the planet, of the understanding of death." [1] Once again, the fascination circles round to a spiritual purpose and total worldview; in

1. George Leonard, *The Ultimate Athlete* (New York: Viking Press, 1975). There are some striking anticipations of this new vision of Western athletics in Paul Goodman's early stories and novels, where long-distance running, handball, cycling, basketball appear as rapturous activities. In Goodman's case, the sensibility grew out of his Taoist-Gestalt worldview, which gracefully blended organism and psyche.

this case, it is running or rock climbing, skin diving or "tennis flow" that brings us to what Leonard calls "the game of embodiment that the individual human spirit plays on the magic playground of this planet."

A little further along the consciousness circuit and we arrive at the mystic and organic food cults of the day. Here, as with the etherealized healing and exercise techniques, the interest assumes for its devotees an intensity and metaphysical elaboration which makes it far more than a matter of physical health. The occult harmony of the nutriments, the ritualized gathering and preparation of foods have been transformed into a focus of contemplative attention. Again, by conventional Western standards, this is no place to look for religious experience. Our culture has long since banished such mundane necessities as cooking and eating to the realm of the profane; we are wholly without a nutritional mystique. But the "wholefood freaks" among us have learned to respect the homely wisdom of the Upanishads: "First know food. From food all things are born, by food they live, toward food they move, into food they return." For them, the interplay of yin and yang in the daily diet has become a mandala of the kitchen sink, a sutra of the supper table.

And the spectrum of contemporary spiritual expedients, having run its wayward course through shiatsu massage workshops and macrobiotic food stores, reaches out toward more unlikely fascinations still. Even Western science and technology—or some zany interpretations thereof—fall within its sprawling diversity. There is a restless breed among the younger physicists especially who are pressing the wild, theoretical frontiers of their science toward strange, rhapsodic cosmologies. Thus, the "visionary physicist" Jack Sarfatti, starting from the idea that "matter is gravitationally self-trapped light" and drawing heavily on the theories of John A. Wheeler, David Bohm, and Richard Feynman, hazards the possibility that "reality is composed of an infinite number of coexisting universes" which can be freely "interpenetrated" by altered states of consciousness. From there, it is apparently only a short step to collaborating on an "explanation of the unexplainable" with Bob Toben, a student of Carlo Suarès, "world authority on the energy code of the Qabala." [2] And soon enough, astral projection, Kirlian photography, psychometry are before us as very nearly established facts of nature. The style is hardly my own, and I surely cannot vouch for the use that is made in these rarefied circles of concepts like "wormholes" and "biogravitational fields," "negative mass" and "superspace" (I suspect it is all scandalously farfetched.), but I can recognize the underlying intention of the effort. It is to mentalize matter and to portray physical nature as the plaything of the mystic powers. In short, it is religious aspiration redecorating a prison cell called "the scientific world view" to its own perverse tastes.

The technician's version of "visionary physics" is, I suppose, the media metaphysics of Marshall McLuhan and John Brockman. Here electronic communications and information transfer systems become the sacramental powers of a psychotronic mysticism whose object is clearly to boggle the mind with paradox and astonishment. Angular analogies and "interfaces" are worked out between advaita vedanta and cybernetic technology; physical reality dissolves into the free play of consciousness, and the universe becomes the theater of all possibilities.

At last, if we follow the consciousness circuit to its pop culture extreme, we find science fiction, acid rock, and "sword-and-sorcery" romances enjoying an almost reverential re-

2. Sarfatti contributes the academic and theoretical back-up to Toben's rather too cute and cartoony little work, *Space-Time and Beyond* (New York, Dutton Paperback, 1975). But the book's bibliography and Sarfatti's "scientific commentary" are a serviceable guide to "visionary physics." Sarfatti, whose academic credentials and publications are unexceptionable, is head of the Physics/Consciousness Research Group in San Francisco.

spect as doors of extraordinary perception. The objects are vulgar enough, but the appreciation they attract can take on a remarkable, if scarcely warranted sophistication—much like the admiring attention pop and camp artists have lavished on soup cans and movie queens, recognizing in them the debased icons and idols of our society. Even comic book fantasy is concocted and consumed these days with a self-conscious eye for its underlying mythic contours. Surely both the creators and the readers of popular comics like *Conan the Barbarian, King Kull,* and *The New Gods* have learned more than a little about the plasticity of mythological archetypes from Joseph Campbell's *Hero With a Thousand Faces.*

Or consider the quasi-religious pretensions that surround another of the very popular comics, *Doctor Strange, Master of the Mystic Arts:*

Once he was a man like most others—a worldly. man, seduced and jaded by *material* things. But then he discovered the *separate* reality, where *sorcery* and *men's souls* shaped the forces that shape our *lives.* In that instant, he was *born again,* to become a man like no other—a man who left us *behind,* as he strove to stand against the unseen subtle perils hovering *thick* and *black* around our fragile existence.

Low-grade stuff, admittedly, but typical of the style that permeates the score or more of psychedelic-spook comics which have lately become cult objects among college-age readers.

Unconventional medicine and etherealized healing—mystic food cults—visionary physics—comic strip fantasies of "separate realities"—meditative athletics: if we fail to see the essentially religious impulse that animates these fascinations, then not only does much of the culture around us lose its integrated meaning, but we overlook a commanding passion of the time as well. What is it about these and countless similar preoccupations that makes them "religious"? It is the role they play in drawing *meaning* out of *strangeness*—personal meaning out of cosmic strangeness. All are valued for the access they give to uncanny realities, regions of experience that transcend conventional science, standard intellect; each is surrounded by that special luminosity which only belongs to an ultimate human concern. The choices may be banal, the efforts flawed, but the intention is clear. We see bursts and flashes of visionary energy at work here, struggling to break the grip of time and matter, and to find, beyond the world they govern, a universe responsive to our human presence. Of course, the need has always been there. But now, see in how many ways it publicly and unblushingly declares its existence.

The Whole Holy Works

The chart that follows on the next few pages offers some idea of the many portals, grand and humble, through which people currently invite the experience of transcendence—or some fleeting glimmer of it—into their lives. Admittedly, my inventory hardly does justice to the number and variety of means that cater to the spiritual need of the time, nor to their exuberantly eclectic and overlapping tendencies. For the fact is, one would be hard put to name anything that somebody somewhere has not ritualized or philosophized into an introspective exercise, if not a full-blown spiritual discipline. Certainly in the San Francisco Bay Area, where I live—one of the hothouses of the consciousness circuit—it has become impossible to keep track of the quasi-religious and therapeutic trips people are "into"—and out of, very nearly by the week. Fickleness and superficiality are obvious vices of so transient a scene. But even the vices describe a significant sensibility, one which asserts, by the range and easy changefulness of its choices, that just possibly anything and everything

AQUARIAN

JUDEO-CHRISTIAN REVIVALS

New Pentecostalism (Jesus Freak sects
and communes)
Charismatic congregations in
mainstream churches
Right On (Christian Liberation Front
journal)

Havurot movement ("Jewish
counterculture")
Chabad campus houses
House of Love and Prayer, San Francisco
(Rabbi Shlomo Carlebach)

EASTERN RELIGIONS

Zen
Tibetan Buddhism
Tantrism
Yoga
Sufism
Subud
Baha'i
Taoist nature mysticism
I Ching

Personal Gurus and Mass
Movement Swamis:
Krishna Consciousness (Swami
Bhaktivedanta)
Transcendental Meditation
(Maharishi Mahesh Yogi)
Divine Light Mission (Maharaj Ji)
Healthy-Happy-Holy Organization:
3HO (Yogi Bhajan)
Integral Yoga (Swami Satchidananda)
Ananda Marga Yoga
Meher Baba
Naropa Institute-Dharmadatu
(Chögyam Trungpa)
Nyingmapa Institute (Tarthang Tulku)
Sri Aurobindo
Swami Muktananda
Sri Chinmoy
Kirpal Singh
Guru Bawa
Bhagwan Shree Rajneesh
Eknath Easwaran
Gopi Krishna
Master Subramuniya
Buba Free John
Baba Ram Dass

AQUARIAN FRONTIER:
POINTS OF ENTRY

FRONTIER

ESOTERIC STUDIES

Studies in Comparative Religion
(journal of traditionalist studies)

Esoteric Groups
Theosophy
Anthroposophy (Rudolf Steiner)
Gurdjieff—Ouspensky—J. G. Bennett
A. A. Bailey

Kabbalism
Astrology
"Humanistic Astrology" (Dane Rudhyar)
Alchemy
Tarot
Magic
Geomancy (leys and power points)
Occult histories (lore of Atlantis, etc.)

EUPSYCHIAN THERAPIES

Jungian psychiatry
Gestalt
Psychosynthesis
Primal Therapy
Arica
Erhard Seminar Training
Centering
Humanistic Psychology
Transpersonal Psychology
Logotherapy
Synanon games
Silva Mind Control
Mind Dynamics (A. Everett)

ETHEREALIZED HEALING

Integral Healing
Acupuncture
Polarity Therapy
Autogenesis
Homeopathy
Naturopathy
Hypnotherapy
Aura readings
Psychic surgery
Iridology
Yoga asana therapy
Radionics
Herbology
Gem and flower therapies
Reflexology
Biotonics
Medical astrology
Aletheia Foundation (Jack Schwarz)

BODY THERAPIES

Sensory Awareness
Structural Integration (Rolfing)
Structural Patterning
Bioenergetics
Orgonomy
Alexander method
Feldenkreis method
Well-body work
Massage
Somatology
T'ai chi ch'uan, aikido, other oriental
exercise and martial arts
meditations
Therapeutic athletics (Esalen
Sports Center)

A Q U A R I A N

NEO-PRIMITIVISM AND PAGANISM

Philosophical mythology (Jung, Eliade,
  J. Campbell)
Sorcery and shamanism (Don Juan,
  Rolling Thunder)
Voluntary primitivism as a lifestyle
Adaptations of primitive lore and ritual

ORGANICISM

Ecological mysticism
Natural foods cults
Macrobiotics
Organic husbandry
Biorhythms
Fruitarian dieting

WILD SCIENCE

Altered states of consciousness
Biofeedback
ESP and parapsychology research
Dream research
Psychometry
Psychedelic research
Kirlian photography
Schmidt machines
Life Fields
Split-brain research
Age-regression hypnosis
Morphic science (L. L. Whyte)
Visionary physics (universe as
  consciousness)
Psychoenergetic systems
Parapsychic/physics interfacing
Thanatology (death and dying)
Synergistics (metaphysical
  geometry: B. Fuller)
Research Centers:
  Institute of Noetic Sciences
    (Edgar Mitchell)
  Foundation for Mind Research:
    Mind Games (Jean Houston)
  Kundalini Research Institute
  Central Premonitions Agency

AQUARIAN FRONTIER:
POINTS OF ENTRY (CONTINUED)

FRONTIER

PSYCHICS, SPIRITUALISTS, OCCULT GROUPS

Edgar Cayce
Uri Geller
Eckankar
Stele Group
One World Family
Summit Lighthouse
Church of Divine Man
The Process
Pyramidology
Gnostica (occult journal)

PSYCHOTRONICS

Neural Cybernetics
Media mysticism and electroneuronics
Drug and electronic manipulations
   of the brain

POP CULTURE

Science fiction
Metaphysical fantasy
UFO cults and study groups
Sword and sorcery romance (Tolkien,
   Peake, Cabell, Lovecraft)
Comic book fantasy
Acid rock
Light shows and multi-media
   spectaculars
Cinematic turn ons
Dope and unstructured mind-blowing

can give wings to the questing mind. After all, once we see there is a Zen of archery and of basketball, then why not a yoga of "chew-fifty-times-before-swallowing" or sadhanas of organic gardening and honest craftsmanship? What limits can we assign to the illumination of the commonplace? "One should regard oneself and all that is visible as a divine mandala," Lama Govinda has said. And there are clearly those who are trying for all they are worth to achieve just such a divinization of nature and all human conduct. It is something we have picked up from the Zen-Taoist sensibility, this eye for the contemplative and ritual possibilities of ordinary experience. We learn from the oriental art of flower arranging and the tea ceremony how to ponder the whole holy works about us as a storehouse of enlightening moments. The new psychotherapies especially have been prompt in recognizing how this homely mysticism of the here-and-now can be used to center and mend the personality. That is, in fact, what makes them, in Abraham Maslow's phrase, *eupsychian* therapies: therapies of health and growth.

How does one speak of a search that chooses so many diverse vehicles to carry it forward? We have no word in our language to name it; we would have to concoct some impossible linguistic hybrid: "psycho-mystico-parascientific-spiritual-therapeutic . . ." Having no better term to corral so wide and wild a range of pursuits, let us borrow an allusion from the popular mythology of the day and call the scene as a whole, with all its paths both straight and twisted, the "Aquarian frontier," the subtle landscape and open field of contemporary spiritual adventure.

Precisely when we entered the Aquarian era, or whether we have as yet, is a matter of debate among the astrologers; but all agree that it is a phase characterized by high moral idealism and an inviting openness to visionary experience: a naïvely questing stage of history whose advent brings with it a startling shift of the sensibilities toward deep subjectivity. Yet, as introspective an era as this promises to be, it is also a time that reaches out toward more intimate forms of community and toward ambitious cultural synthesis. Dane Rudhyar reminds us that the Aquarian is an oceanic phase; and, like the intermingling seas of the world that are in reality one connective tissue binding all continents and islands together, its supreme purpose is planetary unity, the "symposium of the whole."

But perhaps the most widely appealing quality of the Aquarian sign is the hope it offers as water-bearer to a parched and dying culture. Aquarius, the bringer of water, an emblem of life in the midst of a wasteland. Or such is the promise of the frontier before us, though we must bear in mind that, where fertility is not matched by careful cultivation, it yields no livable human habitat, but instead the deadly luxuriance of swamp or jungle.

And where better to begin examining that promise than with an image of the frontier that displays both its vitality and its riotous promiscuity. For it is spiritual exploration as a *mass* phenomenon that provides our subject matter. So I choose an event which draws out the troubling ambiguity of the rising and unruly public appetite for psychic renewal: The Kohoutek Celebration of Consciousness, held in San Francisco during late January 1974.

"A Parliament of Monsters"

Does one call such an event a "conference" . . . "convention" . . . "rally"? "Carnival" might not be wide of the mark. In fact, the spectacle brought to mind Wordsworth's description of St. Bartholomew's Fair.

> All movables of wonder, from all parts
> Are here . . .
> All out-o'-the-way, far-fetched, perverted things,

> All freaks of nature, all Promethean thoughts
> Of man, his dulness, madness, and their feats
> All jumbled up together, to compose
> A Parliament of Monsters.

The Celebration's organizers sought to describe the affair in several ways: "a ritual for 8000" . . . "a gathering that in itself makes visible the ways in which disciplines are coming together in their mutual quest for a higher human consciousness" . . . "an alchemical mix for transmitting and refining our individual approaches to a new art/science of consciousness."

To achieve that mix, the Celebration brought together several score of the country's leading gurus, therapists, consciousness researchers, "psychic facilitators," and surrounded them with all the occult and religious groups for whom floor space could be found in the San Francisco Civic Auditorium. From this heady brew of resources, a weekend program of workshops, panel discussions, films, and sundry entertainments was assembled for public examination. For the price of admission ($5.00) people were invited to spend up to fourteen hours each day wandering among the exhibition booths, sampling the occult goods and exotica on display, absorbing ethereal vibrations. On the central platform, a steady schedule of dialogues, dance pieces, and lectures unfolded. The West Coast "movement artist" Ann Halprin appeared at several points to improvise mass rituals. Mystic musicians (Francisco Lupica and the Cosmic Beam, the Gravity Adjustors Expansion Band, Amazing Grace, the Sufi Choir, the Mantric Sun Band) were on hand to perform. The One World Family of the Messiah's World Crusade took charge of the natural foods concession, featuring organic brown rice under rich garbanzo gravy, whole wheat sunflower seed bread, and herbal tea.

As for the disciplines and experiences presented: they

made up a stew nearly too miscellaneous for digestion. Upon entering, one received a flyer promising a wealth of astonishments. It read:

In Fifty Rooms, Experience: Aikido, Kirlian Photography, Herbs, Psychosynthesis, Holography, Biorhythms, Chanting, Massage, Polarity Therapy, Gestalt, Dreamwork, Biofeedback, Mantras, Mandalas, Bio-energetics, Astrology, Yogas, T'ai Chi Ch'uan, and many more.

"And many more" turned out to be much more indeed. If one searched out all the booths and all the workshops in all the corners of the auditorium, one came upon Palmistry and the Tarot, Witches' Rune Sticks, Practical Divination, Manifest Wisdom of the Great Pyramid, Cosmogenic Art ("The Out-of-this-world Art Form that springs from within you"), the Life of the Maharaj Ji on film, BioMagnetic Psychles for Insight, Past Life Readings, UFO Research, Electro-Acupuncture, Chaotic Meditation, Tantric Healing, Shiatsu Healing, Parapsychology for the Business Executive, ESP Teaching Machines, Orgasmic Union, Pathway Vibrations, Karma Cleaning, Universal Bio-imagery. The Intensive Group Workshop promised "a smorgasbord of group experiences, using trust, love, sensuality, anger, and like that." One Psychic Development Counselor advised all comers to "Expect A Miracle Today, For There Will Be Many"; another featured the "Self-Liberation Karma-Buster," made of genuine plastic and selling for only one dollar. The Summit Lighthouse, offering "a journey to the sun in a chariot of fire," invited participants to "travel with the Archangel of the Violet Flame Holy Amethyst to the sun behind the sun. To be preceded by meditation on the music of the spheres with full color slides of the cosmos."

The greatest show on earth . . .

Those who attended—and there were thousands—came with the gaiety and playful credulity of children at the circus, eager to believe and wonder at all they saw. The spirit of the moment was high good humor, with everybody trying hard to participate and to generate a happy astral awareness. People danced in the aisles and passed out flowers. Incense and chanting filled the air. The gurus who showed up spoke cryptically of apocalyptic possibilities and evolutionary breakthroughs. Nobody questioned anybody's visions or convictions. On the way to the men's room, one was offered joss sticks by Krishna-Conscious young people who lined the corridors with drumming and mantras.

Such an event would have been remarkable enough if it stood by itself as a unique occasion. But the Celebration of Consciousness is only one in a lengthy series of mystic pleasure fairs that have been staged in Northern California since the spring of 1969. Its predecessors and successors include:

The Celestial Synapse, San Francisco, June 1969
The First Whole Earth Festival, Davis, June 1970
The Omniversal Symposium, Sonoma, September 1972
The Awareness Festival, San Francisco, September 1973
The Convocation of the Autumnal Equinox, San Francisco, September 1973
The Participative Symposium on Integration for Personal Autonomy, Berkeley, March 1974
The First Annual Psychic Science and Arts Fair, San Jose, July 1974
The Cosmic Mass and Celebration, San Francisco, August 1974
The Healing Celebration of the Autumnal Equinox, Santa Cruz, September 1974
The Dharma Festival, Berkeley, October 1974
The Gala-Galactic Extravaganza, Berkeley, August 1975

In each case, the effort has been to create a spontaneous congregation of expanded consciousness; each event has tried to gratify the same widespread, omnivorous craving to experience the Extraordinary. After so many experiments along the wild-minded Californian margins of postindustrial America, how long before the Aquarian apocalypse assembles its elect in the first planetary Congress of Wonders? Doubtless such a Woodstock of supernatural splendors is close at hand. Perhaps, even now, some bright, young "impresario of consciousness" (as one of the Celebration's organizers called himself) is leasing a sacred grove or Indian burial ground for the event, beaming clairvoyant invitations to our sister galaxies, negotiating film rights with Uri Geller and Father Ely, Finbarr Dolan and the certified, original Don Juan of the Separate Reality.

But I am being too flippant—a temptation it is hard to avoid in the face of occasions that so blithely mingle the wisdom of the ages with unabashed showmanship. Yet, for all the P. T. Barnum philistinism of its style, it would be a severe mistake to underrate the importance of the Celebration. Behind it, like a diffuse and many-colored light which the event has drawn suddenly into sharp focus, stands the freewheeling religiosity that has been an essential feature of countercultural dissent in America for the past generation. In the beginning was the hallucinogenic obsession and sheer, infantile make-believe. But how far we have come since the callow days of LSD and Tolkien's hobbits. The taste for psychedelic delights has reached out to salvage a hundred occult and mystic disciplines, both traditional and experimental; the fairy tale reveries have assimilated (often with marvelous indiscrimination) all the myth and lore of the world. A glance at publishers' lists of the last ten years quickly reveals what a wealth of literature, research, and scholarship has been mined along the Aquarian frontier. Esoteric classics and metaphysical treatises that have circulated in the shadows since they left their authors' hands— Eliphas Lèvi, René Guénon, Dane Rudhyar, Gurdjieff, Ous-

pensky, A. E. Waite, Frithjof Schuon, René Daumal—
now appear in crisp, new paperback editions that find a
greater audience in one year of college course adoptions
than in their entire previous history. Religious studies
courses have become a lively new feature of the university campuses. Major centers of Zen Buddhist, Tibetan
Buddhist, and yoga studies have been transplanted to America by emigré lamas and gurus. Research on psychedelic
experience, meditation, parapsychology, unconventional
medicine is rapidly growing into a new academic profession devoted to the study of "altered states of consciousness"
in the broadest sense. Eupsychian therapies of self-exploration have become a national pastime of the dislocated American middle class and (very nearly) a major service industry.
Religious and psychiatric communities have blossomed into
new, quasi-monastic, family forms. And, most obviously, a
menagerie of ecstatic and meditative religious movements—
the true-believing hosts of Lord Krishna and the pentecostal Christ, of the Maharishi and the Maharaj Ji—has taken
to the streets and the marketplace to recruit the worst casualties among our overdosed and bummed-out youth.

Who would have predicted that so much cultural action
would be spun off by an interest that once looked, for all the
world, like mere childish self-indulgence? Yet it was a generation of mind-blown hippies and flower children, walking
through the world as if with eyes on some other reality, who
opened the Aquarian frontier and made it a visible social
fact, a live option for the millions. They popularized the
mystically distracted style of life; they provided the audience—or, if you will, the market—for the gurus and therapists of the day. Above all, it was these restless and disaffiliated young who publicized the possibility of radical
self-transformation. Playfully casting off their middle-class,
suburban identities, they became Zen Buddhists, American
Indians, Hindu sadhakas, witches and warlocks, gypsies of

the open road, street freaks (and, along other lines, people's warriors and enemies of the state). By the act, whimsical and amateurish as it may have been, they raised the image of *rebirth* like a banner of the times. And rebirth, the chance to expunge the past and begin all over again as an innocent and autonomous soul, is the crying need that now ruthlessly drives more and more people toward harrowing therapies and agonized conversion experiences along the Aquarian frontier.

The Traumas of Rebirth

There is a line of thought among radically inclined psychotherapists which treats all psychic distress at the upper levels of capitalist society as the result of underlying social injustice. From this viewpoint, the anxieties of the affluent are essentially symptoms of political guilt: guilt born of unjust enrichment, of invidious status exploited from the poor.

I believe this is true; but it is true as the special (and highly compounded) bourgeois case of a greater human need, one which remains with us even in a social order unblemished by exploitation. And that is our need to become *serious* human beings, people who grow by virtue of having struggled in the solitude of the heart to find both moral dignity and personal meaning. It is our need to live deeply, to take life in our hands, to weigh and feel it, to give it deliberate shape—our *own* shape, the shape of our peculiar experience. Debased consumer fantasies, selfish acquisitiveness, the careerist rat race, the corruptions of elitism: these are surely among the forces that thwart our need to grow. But there are other ways our lives can be made characterless, trivial, alienated—perhaps even by a "consciousness raising" which aims to make us socially responsible but never personally *interesting*. Or perhaps by all forms of economic security which infantilize us, by all affluence which distracts us from the risks of authenticity, by all social duty which would keep us from being alone with our personal destiny.

Wherever that happens, we know the guilt of having lived below the level of our potentialities. We experience failure, and begin casting about for the chance to suffer and make sacrifice for the sake of new life, for the chance to pass through the ordeal of the fiery furnace where the personality is fiercely purified of its old lies and unworthy loves, melted down to the core, made ready to be reshaped to its true identity. A clean break and a new beginning. Purgation and renewal. Thus, the young acolytes of Lord Krishna deliver up all their worldly goods, shave their heads, take new names, and dress in saffron robes; Erhard Seminar trainees submit to day-long marathon sessions of personal vituperation aimed at destroying all the defenses of the ego; Gurdjieffian disciples agree to be worked and ragged and ridden to exhaustion; encounter groups seek ever more grueling extremes of bullying and abuse.

Traumas of rebirth, all these. Terrible testings whose purpose is the mercy killing of false and faithless identities. And with each, there comes the taste of transcendence. For to be twice-born is an undoing of the past, and, to that extent, a victory over time and mortality.

More than anything else, it is the search for rebirth that has populated the Aquarian frontier. The therapies, meditations, and spiritual disciplines people find there have become so many ways of exchanging old life for new. Though our culture does not respect the necessity or cater to it, the need to be reborn is among the great constants of our human condition; it is the tidal rhythm of life that keeps the personality from stagnating in its accumulated guilt, regret, and failure. But again, the need may assume strange disguises. What, for example, is the real attraction of "Rolfing" (Ida Rolf's Structural Integration), an agonizing, but highly popular body therapy which reorganizes the anatomy by ruthlessly breaking the tough connective tissues (the fascia) which keep the kinks and crooked posture of a lifetime locked into the organism? On the surface, Rolfing offers nothing more than a tough workout and rapid physical realignment. But at another level the ordeal promises to root out the fears, anxieties, rigidities which have been warping our bodies since babyhood. It promises to restore our pristine physique; and what is that but a good bruising way to expunge the sorry past and give back our lost innocence? A second life, won at the expense of much therapeutic punishment.

But not everything that promises rebirth brings the experience of transcendence. There are many bad, secular substitutes on the scene, as well as many forms of "spiritual fascism" which only exploit our need, rather than fulfilling it. Consider, for example, the case of Patricia Hearst, making her torturous way into the ranks of the Symbionese Liberation Army in 1974: certainly the most startling public example we have had of the rebirth experience in modern America. First, the ordeal of her kidnapping and lonely captivity; then, in rapid succession, her passionate conversion to the cause of her captors, the ugly denunciation of parents, friends, lover, and her entire earlier life, the adoption of a new name (Tania), a new costume (Fidelista fatigues), a new tonsure of the hair, a new vocabulary (compounded of Maoist clichés and "people's" obscenities); finally, her initiation into the radically new lifestyle of hard-bitten, gun-toting urban guerrilla, consummated by the performance of a hazardous initiation rite (a daring bank robbery). It is an absolute paradigm of religious rebirth—or perhaps one should call it an absolute parody, since, in this instance, it is the religion of violent revolution and blood vengeance that consumes the life of its disciple.

The example should remind us that people filled with enough self-loathing and desperation are apt to seize upon anything that offers them an explosive catharsis and a decisive break with the past—and to love the hand that drives them to their ordeal. Every totalitarian mass movement of modern times has borrowed and perverted the psychology of rebirth for its own purposes. It guarantees us no humane result that people come flocking to the Aquarian frontier crying out to be born again—not unless the experiences of renewal they find there prove to be ethical as well as ecstatic.

An Ecology of the Spirit

There are those, I know, whose allegiance to social conscience and hard-edged intellect has never allowed them to see the religiosity of the American counterculture as anything but malaise: a soft, insinuative decadence oozing in from the crumbling edges of an expiring capitalist society. "The counterculture is doomed," the activist anthropologist Marvin Harris argues, "so long as everyone is doing his own thing. Head trips and meditation, a million chanting messiahs and love-ins will not appreciably affect . . . material conditions. . . . It feels good while it changes nothing." [3]

An old and honorable lament, and by now a truism. Who would deny that religion has been used both to disguise apathy and to shield privilege? Pie-in-the-sky making up for the poor crust on the table. Contemplation of one's navel as a displacement of political resignation. Of course. But what to do with those—the conventional radicals, the secular humanists one and all—who cannot see how much more is to the matter than this long, weary history of treachery and abuse? Have they never, not even for an instant, felt the visionary energy take heat within them, if only, as with most of us, like fire in a mine shaft, more smoke than flame? If the experience has never been theirs, nothing I say here will prove its reality. Let me only offer them a proposition to challenge the sufficiency of their indignation. For I argue

3. Marvin Harris interviewed in *Psychology Today*, January 1975, p. 69.

that our exploitation at the hands of the industrial elites who so distort our lives has been worse by far than our radical forebears realized. We have been cheated of *more* than bread and justice, and the crime is no less a crime because our exploiters themselves suffer from the same privation of spirit as we.

We know by now, do we not, that oppression does worse than punish the flesh? It can darken the mind, shrivel the sensibilities, narrow the personality until there remains no room in our lives even for minimal self-respect. If the sick priorities and benighted leadership that have blighted industrial society for the past two centuries can do so much to diminish the size of our lives, why not more and worse? The privation which our spiritual needs have suffered in Western society has been a facet of the same "industrial necessity" that has warped all our ideals and institutions. Indeed, the hard discipline of primitive accumulation and economic growth is even less compatible with transcendent awareness than with democratic politics. If, then, we deny the realities to which religious aspiration reaches out, are we any better than blind men denying the light?

Social justice has for so long been the substance of our politics that we easily forget the needs that lie still deeper down—so much deeper that to speak of them at all may seem to leave behind political discourse entirely for psychotherapy or spiritual counseling. But, then, have we not learned that politics is a Chinese puzzle box: need within need within need, issue within issue within issue? Inside capitalist industrial development and the invidious hardships of its steep ascent, find liberal reform. Inside liberal reform, social democracy. Inside social democracy, personal fulfillment. Inside personal fulfillment, the transpersonal quest.

We have been called, in some traditions long excluded from the Western cultural mainstream, children of the light, imprisoned angels, amnesiac gods, beings of infinite and immortal magnificence. Any politics that degrades our human potentiality below that sublime level is alienation's child still, however powerful its ethical passions. *Liberty, Equality, Fraternity* . . . but what are all these without *sanity?*—the health and wholeness of the soul? That is what we must look to see in the impetuous search for "new consciousness" that now arises along the Aquarian frontier: an effort to find the dimensions of sanity. And if that project has become for some an all-consuming preoccupation, perhaps they are more right than wrong. Perhaps their fascination with exploring and experiencing the Extraordinary stems from a valid urgency. For I too suspect that what is at stake in this too-troubled generation—as much as our physical survival on this planet—is the survival of the visionary energies. By which I mean the survival of our ability to perceive, if only dimly, the life-enhancing truths to which myth, ritual, sacramental symbol, contemplative object, magic rite, the natural wonders, and ecstatic communion bear witness. Or perhaps the powers I speak of would be better recognized by the derisive names bestowed on them through twenty centuries of grinding Christian orthodoxy and three hundred years of crusading scientific intellect: superstition, idolatry, heresy, animist primitivism, unreason, mystical nonsense, plain madness.

Somewhere in that slanderous slag heap of exiled and forbidden human capacities, like diamonds in a dunghill, lie prizes of vision that until recent years only a few in the modern West have dared to search out, and fewer still have reclaimed with any sure and graceful discrimination. How else to understand the reckless contemporary need to try anything—*anything*—that will jar, shift, trick, trap, batter, entice the mind into states of strange awareness: any pill, any punishing therapy, any recourse to psychosis or fantasy? A rare and irreplaceable resource of the spirit has been slipping steadily away from us, vanishing from the Earth under pressure of industrial necessity and technocratic rationality. *Deus absconditus . . .* and we reach out in desperate longing toward any least glimmer of the failing light, striving to draw back the retreating god.

That is why these ancient and unlikely life signs now come flooding into our culture—mandalas and god's eyes, Egyptian ankhs and the Tarot deck, intruding themselves among the supercomputers and the moon rockets. Relics of the departed god. Even in their most tawdry and juvenile versions, they tease our diminished mind with memories of a lost knowledge, of gnosis, which is knowledge of the sacred attained at the threshold of ecstasy. At the climactic summit of industrial civilization, just as all the human and environmental systems on which that civilization rests begin to crumble beneath us, we discover that there is also an ecology of the psychic environment which must be trusted, like all life support systems, to find its way to balance and compensation. Either that, or the human personality, which is the richest of all the world's gardens, becomes waste and toxic. Already, when we peer most candidly into our urban-industrial future, it is a harvest of nightmares we see springing from our poisoned soils: death camps and nuclear Armageddons, Beckett's Endgame and Huxley's Brave New World, the terrors of Soylent Green and the Clockwork Orange.

To create a healthy ecology of the spirit, then: that is the task we confront on the Aquarian frontier. To let the wheel turn, the cycle come round—through healing darkness back into morning. Mind, having reached the end of its tether, has no choice but to begin its journey to the underworld, struggling amid the annihilating despair to recapture its visionary core.

2

God Between the Carnival
and the Computer

The inward voyage is the emergent sign of our times. But, as the Celebration of Consciousness suggests, not all who voyage are homeward bound. Some are out for larks, some are conducting lucrative pleasure cruises, some are stranded on shoals far from the oceanic deeps, some are merely researching the surface currents, some are fast vanishing into the maelstrom which destroys mind, soul, and spirit all.

How do any of us, children of a culture so long estranged from the visionary talents of the human animal, pick our way through this turgid era of psychic adventurism with any confidence that we will choose wisely among the proliferating options? Nothing is guaranteed about our trek into the Aquarian frontier, least of all the capacity for regaining our spiritual balance without many a trial and hazardous error. For all the urgency of our need, there is too much too suddenly before us, and in these shadowy areas of human experience we are perhaps the most amateurish people in the history of the world.

Until no more than a generation ago, it was enough—it was all that any self-respecting intellectual dared do—to call for simple tolerance and openness toward religious tradition. In the withering skeptical climate that has prevailed in our intellectual mainstream since the Age of Reason, even the best clerical minds learned to tailor their commitments to one or another of the current liberal and humanist fashions: natural religion, Romanticism, the social gospel, psychoanalysis, existentialism, death-of-God. Thanks to this strategy of cunning adaptation and protective coloring, the churches have not only been poorly armed against their agnostic opposition, but they have surrendered their position as an independent cultural force, and so have had nothing to contribute to the exploration of the Aquarian frontier. The last thing it seems our clergy ever expected to witness was the reemergence of religion in the modern world as an autonomous, culture-redeeming experience of gnosis and prophecy. In another respect, too, the mainstream churches have paid a high price for their kowtowing to secular intellect. They have seen their bewildered and needy flocks become easy prey to pentecostal raiders, whose charismatic appeal has made their tight-minded sects the fastest-growing of all contemporary Christian congregations—and the least capable of helping create an integrated planetary culture.

Even the mavericks of the modern West who have been willing to rub against the positivist grain—the artists working mythic and mystic veins of inspiration, the offbeat philosophers and psychiatrists—have often been intellectually fastidious in the extreme when treading the forbidden ground. T. S. Eliot, loading his *Waste Land* with a cargo of unfamiliar religious curiosities, made sure to ballast the poem with leaden scholarly authority. Jung, who labored to keep his work in religious psychology as academically reputable as possible, buried his deepest spiritual explorations in private printings and in his still unpublished Red Book, hiding them away like guilty secrets. The same caution colors the groundbreaking efforts of Frazer and Yeats and Joyce to pry open the Western mind to the invaluable insights of myth, ritual, symbol. In all of them there is the same taste for learned allusion and exhaustive study, as if one needed the license of heavy scholarship to traffic in such exotic wares. Even then, they made their point obliquely, unobtrusively—even pedantically.

Hard enough in those days to get one's foot in the door of the cultural establishment with any viewpoint that even hinted at the insufficiency of secular intellect. Now, in just the last ten years, the doors are off their hinges; and, with the house overrun by the varieties of religious experience, how quaint the caution of a generation ago looks. As quaint as the first timid appeals for sexual liberalization in our grandparents' day now seem to a society that finds itself drowning in permissiveness and pornography—a similarity which is more than fortuitous. We have, in fact, much to learn from the strikingly parallel courses sexual and religious emancipation have followed in our culture. Both have burst into a vacuous and dizzying freedom after long incarceration —like prisoners escaped into unfamiliar terrain with no idea which way to turn. And so both have run to the same dangerous confusions.

The public unfettering of sexuality can now be seen as one of the earliest phases of "consciousness expansion" in the Western world—a pathway into the Aquarian frontier still being pressed forward by the new physical therapies and body yogas. When Freud began his historic excavation of the unconscious, it was inhibited sex he uncovered at the first level of repression—not far below the surface of the mind, even though, for some two generations, the erotic energies were widely taken to be the bedrock of the personality. The sexual permissiveness that followed upon his work has brought many obvious benefits with it. But still, half a century after Freud, our love life finds itself torn between the sexologist's clinic and the pornbroker's orgy.

On the one hand, we must have "the joy of sex" sanctioned, if not prescribed for us, by certified, scientific authority on the basis of expert opinion and mass surveys; and with the sanction, we want all the pills, therapies, and research that will guarantee our doing exactly the "right" thing in the "normal" way. On the other hand, we identify indiscriminate sexual release as "liberation," forgetting that what has been repressed may return to us severely warped by its experience of long imprisonment, and so require careful reentry into the light of day. Thus, we find the casual frivolity of *Playboy* magazine masquerading as "sexual freedom." Or, worse still, the very hard-core obscenity that is the prime symptom of repression presents itself on film and in print as superlative eroticism—and with such a self-congratulatory boldness that it has become possible for children of nine and ten to buy in bookshops across the country a proliferating species of "adult" (so-called) comic book, in whose crudely rendered pages they find nightmares of surrealist sadism that make Hieronymus Bosch look, by comparison, like Victorian nursery wallpaper. All the horrors of Grand Guignol and Buchenwald, stalking out of the dungeons of the mind to become our children's literature.

Clearly, neither Masters and Johnson nor the Sexual Freedom League has so far provided the secret of an easy sexual maturity; the one leads to too much sex in the head, the other to too many fierce or fashionable compulsions in the glands.

So too with our religious life in the first blush of its emancipation. It is also being divided between bloodless research and mindless sensationalism. We have trapped God somewhere between the brain physiologist's computer and the carnival funhouse where everything novel and naughty draws the crowd. These are not extremes that yield some convenient middling course; they are false starts, forgiveable perhaps as pioneering efforts, but false starts nonetheless. And if we are ever to settle the Aquarian frontier, we must mark the roads that lead to barren land. The lines are not easy to draw, for they cut across sincere intentions; but the time is at hand to draw them.

Let us take up the two vices one at a time. First, the desiccation of spirituality by research, and second, its trivialization by permissiveness.

Satori and the Encephalograph

At first glance, it seems anomalous that William James, the father of laboratory psychology in America, should also be among the first serious students of mysticism in the modern world. But the paradox is easily resolved. James, with a behaviorist's keen eye for the empirical, simply recognized that religion need not be (as it has for so long been in the Western world) a matter of belief and doctrine. It can also be *experience*, and once seen from that perspective, the insistent theologizing of Christian tradition begins to look like no more than a thin, verbal oil slick floating atop the deep ocean of visionary religion.

James, for his purposes, was willing to work from literary sources and introspective reports. The result was an urbane study that still stands as a modern classic: *The Varieties of Religious Experience.* But once an empirical element in religious and paranormal psychology had been identified, later researchers were bound to demand more controlled and quantifiable data. Which is what J. B. Rhine was after when he opened his now famous laboratory at Duke University for the statistical study of parapsychology in 1932, and so too Thérèse Brosse in the first electrocardiographic study of yoga in 1935. These were the dim beginnings of what has become, within the last fifteen years, one of the most active and popular areas of psychophysiological research. The boundaries of the field have not yet hardened into a well-defined discipline; the profession still lacks a fixed title. It is sometimes called "transpersonal" or "paranormal" psychology"; sometimes "altered states of consciousness"; sometimes "extrasensory perception," "the psychology of consciousness," or, simply, "consciousness research." One television documentary (NBC, April 26, 1974) referred to it as "Wild Science." Trance, hypnosis, yoga, faith healing, acupuncture, meditation, biofeedback, out-of-the-body and psychedelic experience, plant consciousness, split-brain and dream research—Wild Science has brought them all into the laboratory for quantitative examination.

All this may seem a daring and liberating direction for science to take. I suspect that for some researchers the interest is a sincere, perhaps even slightly desperate, attempt to move their professional work closer to religious values for which they feel a deep personal yearning. Undeniably, some in the new profession (I think especially of figures like Gay Luce and Charles Tart) have done an admirable job of opening themselves and their audience to new ranges of experience for whose investigation they feel a strong emotional commitment. But taking the field as a whole, one cannot help but have stubborn reservations about the contribution consciousness research makes to our spiritual awakening. A wild science perhaps; but the wild new wine is being stored away in very old, very dusty bottles. As exotic a subject matter as altered states of consciousness may be, the style of mind that takes up the study is all too conventional. It *has* to be. Because it is rigorous methodology that is meant to dignify the dubious subject matter. In this area of life as in all others in our culture, the goal of science is to objectify and quantify, on the assumption that only in this way can any phenomenon be made "real" and its study reliable. As if we had more to learn from an electroencephalograph than from all the Upanishads put together.

Even where such ethnocentrism is avoided, the very nature of the research works to reduce and distort what it studies. For one thing, the hard focus on measurable, empirical effects isolates what may well be the least important aspect of religious experience, leaving behind the total life discipline which is both its root and its blossom. What remains is a collection of data points: statistical series, pointer readings, computer printouts, graphs and charts. The temptation, then, is to believe that the behavior which has thus been objectively verified is what religious experience is *really* all about, and—further—that it can be appropriated as an end in itself, plucked like a rare flower from the soil that feeds it. The result is a narrow emphasis on special effects and sensations: "peak experiences," "highs," "flashes," and such. Yet even if one wishes to regard ecstasy as the "peak" of religious experience, that summit does not float in midair. It rests upon a tradition and a way of life; one ascends such heights and appreciates their grandeur by a process of initiation that demands learning, commitment, devotion, service, sacrifice. To approach it in any more hasty way is like "scaling" Mount Everest by being landed on its top from a helicopter.

Moreover, the heavy emphasis laboratory research lays upon special effects cannot help but muddle together things that are on distinctly different levels of spiritual importance. The psychically remarkable or paranormal may easily become confused with the authentically religious, simply because both are exotic effects that have been recorded in the laboratory. Yet all well-developed religious traditions make painstaking distinctions in this area. They distinguish profane madness from divine madness, sorcery from enlightenment. Above all, they distinguish between experiences of the *psyche* and of the *spirit*, insisting that the psychological (or subconscious), where any number of occult powers may reside, must not be confounded with the spiritual (or superconscious), where the transcendent impulse resides. So Jesus warned his disciples that many a false prophet could do wonders more amazing than his own; and the Buddha taught that there was finally only one miracle in which the siddhi should be interested, and that was the moment of *paravritti*, the radical turning about of consciousness in the illuminated personality. Similarly, throughout their long history of advanced psychic exploration, the Tibetan adepts have tended to treat paranormal powers with much the same aristocratic condescension with which many mathematicians and theoretical physicists treat "applied science." Does this not suggest that the level of being at which yogis may purge their sinuses and decelerate their metabolism (or even levitate) is distinctly below the level of enlightenment?

There is a wise Zen parable which has much to teach us about the place of paranormal astonishments in the spiritual life. It tells of the Hindu swami and the Zen master who came to a wide river. The swami proceeded to walk across the surface of the water, at which the Zen master became very distressed and demanded he come back and cross the river in a manner becoming to his exalted calling. The swami returned to the bank. Then both men walked downstream until they came to a shallow place . . . where they waded across.

All this is not to deny that there is much of value in the psychological and neurophysiological spin-off we collect from the experience of yogis, mystics, and psychics. Both our medical science and our psychotherapy need all the help they can get in expanding their vision of human potentialities; and there is no reason why they should not use mystic experience and occult tradition to light the shadowed portions of the personality. The danger sets in where we invert the natural priorities that govern our spiritual growth, placing a higher value on the spectacular by-products or immediately useful effects our science can glean from religious experience than on the visionary insights those effects surround.

Our society has accumulated more scholarly and scientific data about religious and paranormal experience than any people have ever possessed before in history. But too often our research refuses to accept that the healing knowledge we need is a *traditional* knowledge—ours to borrow, not invent. Indeed, it may be a primordial knowledge from which we have been led astray by our very desire to know and to measure objectively . . . as if we searched for truth with an instrument that destroyed what we sought upon contact. In this area of life, what may be required of us is an intelligent deference toward wisdom wherever we find it, even in a past too long regarded by Western society with contempt or condescension. Humility is a quality that bears impressive sanctions in our culture. A celebrated Christian virtue, it has also been championed in modern times by scientists who have insisted that, where our knowledge ends, honest intellectuals are honor-bound to admit their ignorance and say "I don't know." But humility, both Christian and scientific, has an ironically arrogant twist to it. Humility has never prevented Christians from proclaiming that the Logos is wholly and exclusively theirs. So too the humility of scientists has always had a missionary smugness to it. For the implication that lurks behind the scientist's "I don't know" is ". . . and if science doesn't know, nobody knows." Skepticism can be the mirror image of dogmatism, every bit as fanatical and insolent. And both can be knives that sever us from all that sacred tradition has to teach.

The Split-Brain Follies

Within the past few years, we have had before us a blatant instance of this "humble arrogance" that clings to the scientific mind. I refer to the work that has been done in split-brain psychology by Robert Ornstein and others. Let me dwell on it here as an instructive example of how crushingly incompetent science can be in handling metaphysical issues and religious experience—in this case seeking to exaggerate a minor laboratory finding into a total world view, at the expense of endless obfuscation.

At bottom, there is nothing more to be made of the split-brain "breakthrough" than of any localization of functions in the brain. In this case, the much-publicized finding is that the brain is bicameral in its general division of labor, the left hemisphere specializing in language, the right hemisphere in holistic scanning and "intuitive" perception. So fascinated have the split-brain researchers become with their discovery that they almost make us forget the far more remarkable fact, which is that the brain is, at last, a coordinated whole which integrates all its functions continuously, instantly, and spontaneously.[1]

This cerebral division of labor has been nominated by Ornstein and other split-brain enthusiasts as the definitive explanation of all metaphysical and psychological dualities; it has also been presumptuously offered as "proof" that mysticism, meditation, yoga, and aesthetic perception are *important* and even *valid*. Why? Because research has finally given these "intuitive" forms of experience a neurophysiological location in the brain. Only *now*, therefore, do we know they *really* matter and need be taken seriously. So we have ambitious critiques of culture in terms of left/right dominance in the brain. Scientific and technological societies like our own, we are told, are "left hemisphere dominant" or "Aristotelian"; whereas the oriental cultures, with their emphasis on contemplative religions, are "right hemisphere dominant," or "Platonic." The "left side talents" are those of decisiveness, analysis, logic, and articulation. While these have been superbly developed in the West,

---

1. The research was pioneered by R. W. Sperry; see his essay "The Great Cerebral Commissure," *Scientific American,* January 1964. Also see Robert Ornstein, *The Psychology of Consciousness* (San Francisco: W. H. Freeman, 1972); "Right and Left Thinking," *Psychology Today,* May 1973, and Roland Fischer and John Rhead, "The Logical and the Intuitive," *Main Currents in Modern Thought,* November-December 1974.

now "the evidence increasingly favors a generous view of the right half-brain, whose role may be far more important than we know today." But, unfortunately, "we know almost nothing about how the right hemisphere thinks, or how it might be educated." [2] And apparently we are doomed to stay in this state of woeful ignorance until the psychologists report in with further findings that may bolster their mercifully "generous view" of intuitive experience.

What a perfect example this is of the pretentious folly that results when scientists seek to become "empirical" about moral and metaphysical questions. The discussion does not become more precise; it only becomes muddled to the point of disaster, and increasingly more so as we try to replace ethical and value judgments with physical entities and measurable behaviors. To begin with, what is it but semantic legerdemain to identify the visuo-spatial scanning of the right hemisphere with "intuition," then with artistry, then with symbolic understanding, then with magic, myth, ecstasy, satori, beatific union? (Some researchers produce an even more dazzling blur of characteristics, calling the right brain "feminine," "creative," "humanistic," "radical," "countercultural." Start with a sloppy conception of "intuition" and you can finish anywhere you please.) At what expense to the true quality of these many forms of experience do we jam them into the same bin? By what right do people recording brain wave readings related to the perception of space relations or musical tones suddenly begin to expound on "Platonic" and "mystic" virtues? It is commendable, I think, that so many physiologists should want to say a good

2. Maya Pines, *The Brain Changers* (New York: Harcourt Brace Jovanovich, 1974). *"The evidence"* the author speaks of is, of course, laboratory evidence; and the "we" who "know almost nothing about how the right hemisphere thinks" are apparently all intellectually responsible members of a straight-thinking, scientific society. In short, what science doesn't know, nobody knows . . . . least of all the people called artists, visionaries, saints, seers, sages, mystics, etc.

word for mysticism; intriguing that Ornstein should have so strong an interest in the Sufi masters (he sprinkles his writing with quotations from Idries Shah). But how pathetic that all this must be strained through a narrow neurophysiological filter.

Or, again, what sense does it make to speak of a "verbal" ability that is something other than "intuitive" (recalling that "verbal" is left, "intuitive" is right)? In so far as intuition has to do with holistic insight, then we learn, use, and comprehend language "intuitively." Only "computer language" (a term that is in itself a risky metaphorical extension of true speech) is rigorously linear and sequential, a rigid logic of formal bits and pieces. That is why computers are such poor human mimics: they cannot let language bewilder them into philosophy and poetry. When you and I speak, we are expressing whole ideas, global thoughts which pop into the speech-making mind all at once: gestalts of meaning which spontaneously happen into words. If we are Dante or Li Po or the prophet Amos, we can even use language "mystically," "ecstatically," "creatively." And where are we then, on the left side or the right side? Human thought has been plagued by many dualities, but by far the most stultifying is the split-brain researchers left/right.

Whenever we talk about the loss or imbalance of human potentialities, we are talking about repression, not cerebral lateralization. And repression is a quality of the *mind*, not the brain. It proceeds from a distortion of the total personality, dynamically caused by some cultural, political, or economic transformation. We hardly needed the brain physiologists to tell us that our culture suffers from imbalance and repression. Their dubious contribution to the discussion has been to try substituting hemispheric locations for qualities of mind. But once we do that, we no longer have any sensible way of discussing the question at hand.

The very research Ornstein and his colleagues speak from goes on to make clear that (except in brain-damaged individuals or those who have undergone split-brain surgery) all information is continuously exchanged and all behavior coordinated between the brain's hemispheres, in such a way that any so-called dominance of the hemispheral functions has no philosophical meaning. The brain is always in harmony with itself; but there are an infinite number of possible harmonies it can sound—some are sublime, and some are no sane music. To discriminate among these harmonies, however, is a matter of judgment, based on moral and metaphysical commitments that cannot be honestly evaded in any discussion of sanity. There is no talking about the human condition without talking qualities, values, decisions, choices—the attributes of mind.

To play any role in life—whether that of logician or poet, operations analyst or saint—requires the use of a whole, functioning brain. Both Hitler's *Mein Kampf* and Tolstoy's *Confession* are products of whole and coordinated brains working intuitively and discursively at the same time; but these were brains obedient to very different mentalities. Similarly, all the Nazi art produced to serve the Führer's cause was as much the product of "right hemisphere talents" as any art produced by Cézanne or Leonardo—if indeed one can imagine painting being done at all in the absence of analytical ("left side") abilities. The great question is: What shapes the minds that use the coordinated powers of the brain in such widely divergent ways, and how shall we evaluate that use? It does no good to say our education is to blame because it over-emphasizes "left side" talents. *Why* does it do so? In obedience to what values, what interests, what vision of life? It is even less of an answer to speak of an "interhemispheric competition for dominance," as if brains were independent of the personalities that use them; one might just as well "explain" racism as a competition among skin pigments to see which will come out on top. Brain physiology simply gives us no language to discuss repression meaningfully. Or is it being seriously proposed that we will someday be able to identify a sane and balanced personality (or culture) as one which achieves symmetrical left/right alpha-wave readings?

To imagine that current research of this kind will ever provide us with an improved and enriched version of the perennial wisdom is, in reality, to call the stone a loaf of bread and the tainted spring a life-giving oasis. For look what becomes of research that cannot see the whole for its parts, or tell the higher from the lower orders of reality. Already we have the Maharishi's "transcendental meditation" being promoted (on the basis of laboratory findings) as the weary businessman's late afternoon pick-me-up: a sort of yogic martini. And in the Soviet Union, the military is busily at work researching second sight and telepathy as possible adjuncts to its war machine. Who doubts that, at the first solid discovery of parapsychic powers in our laboratories, consciousness research will rapidly slide toward just such corrupted applications? Once we scientize, how long before we merchandise . . . and at last militarize?

In the Shadow of the Great Beast

And, on the other hand, we have the carnival, the psychospiritual free-for-all where anything goes and no holds are barred. In contrast to the consciousness researchers who will admit no realities they cannot reduce to measurement, the carnival is the province of total permissiveness. Here, everything is believed—provided it is forbidden or freakish —and whole traditions of religious wisdom are rapidly exhausted like garish fashions of the day.

If William James is to be credited with founding the scientific study of religious experience, perhaps the figure who stands out as the first religious carny is Aleister Crowley, the self-styled Great Beast of the Apocalypse and Wickedest Man in the World, who became, by design, the most willfully rotten of all the early twentieth-century Decadents.

Although he lived into the 1940s, Crowley belongs from first to last to the cra of Swinburne and Verlaine, Wilde and Péladan. But in his case, the priapism and sadistic obsessions were laced with a powerful fascination for esoteric religiosity. In the one personality we find all the terrible, early confusions of both sexual and spiritual emancipation.

Undoubtedly a man of extraordinary psychic gifts, Crowley fancied himself a connoisseur of all things occult and paranormal. His range and, very occasionally, his depth were impressive: magic, yoga, witchcraft, Tantrism, narcotics hard and soft, alchemy, the Tarot, clairvoyance, demonology, astrology, Kabbalism, the lore of esoteric societies . . . there is little on the consciousness circuit today, other than its scientific research, which Crowley did not sample or claim to have mastered. And yet, there is not a hint about the man of the modesty, gentleness, and simple intelligence which wisdom demands. On the contrary, his life and thought are a picture of mind-bending congestion. At every turn, he mercilessly exploited his talents for sensational self-advertisement and pompous buffoonery. Every tart and mistress in his life had to become the Whore of Babylon; every tawdry seduction had to become an act of ritual magic; every rite and ritual had to be given an orgiastic twist, but never with any sense of the insight which a true Tantric style of eroticism may yield. Everything Crowley turned to had to be pushed to an exhibitionistic extreme. He filed his canines to points and parodied Count Dracula with all the aristocratic ladies he drew under his spell. He immolated cats to rid himself of dysentery. He shocked the world by claiming to have performed human sacrifice a thousand times over. Nothing could temper his towering egotism, and at last whatever he touched became too silly or too terrible for sensitive souls to respect. He was a man intoxicated by sudden, dazzling immersion in powers beyond his intelligent use; in his shambling, lurid career we can see all the giddy vices that come of a premature confabulation with religious and occult tradition.

Imagine the Kohoutek Celebration of Consciousness combined with the Satanic mumbo-jumbo of the Manson family (mercifully and commendably, the one note that was not sounded at the event), and we have the perfect image of the Great Beast's mentality: overcrowded, sensation-hungry, wholly indiscriminate, a vast, seething cauldron that boils up the gristle with the lean, the maggots with the meat. It should so obviously be an image to avoid; and yet it is the shadow of Aleister Crowley, the thrill-seeking apprentice running wild in the sorcerer's palace, that lies heavy upon the carnival, constantly threatening to darken its innocence.

Because religion has for so long been practiced with dogmatic rigidity in our culture, it is inevitable that the Aquarian frontier should be treated as free and unrestricted territory, open to all comers. Again, we are compensating and overcompensating for the distortions of the past, for generations of doctrinal hair-splitting and bloody persecution. On the frontier, the emphasis is upon personal discovery and collective proliferation: everybody does one's thing, and the more the merrier. Such good-natured tolerance is not to be entirely condemned; but, in truth, there is much nonsense and no small amount of opportunism that gets swept into the carnival. What, after all, is one to say of events like the Celebration which purport to put the wisdom of the ages on display in exhibition booths, one right beside the other? To borrow a commercial term that suits the commercialism of such an exposition: where is the "quality control" in all this? In fact, there is none, because the impresarios of the event, while ready to defy conventional intellect, have found no alternative standard of spiritual intelligence which they are ready to affirm. Theirs is not a journey from "here" to "there"; it is a wild rush from "here" toward any place else, and the farther out, the better.

It is just such directionless rambling that brings us to a kind of lumpen occultism which cannot distinguish high tradition from comic strip inanity. The mix is well illustrated in a publication that has lately come my way: *Synergy/Access: A Global Newsletter of Futuristic Communications, Media, and Networking.* The style of the effort might be called Flash Gordon religiosity, compounded in equal parts out of science fiction, parapsychology, Marshall McLuhan, and *Popular Mechanics.* Typical of the "input" on which the magazine draws is "Sensonics Inc.," whose organizer reports in one issue that he has been

zipping around country in van chocked full of mind-tripping electronix, like "Video Chair" and built-in TV studio. Worked with Dr. John Lilly on "bio/video feedback," did video tapes on rolfing and Esalen and generally nosed around far-out spots. Now setting up Florida Growth Center and working on Sensorium. Ask for nifty newsletter, "Boogie On."

The same issue includes a glowing report on Uri Geller's extraterrestrial relations with the UFOs, an item on Chinese oracle bones which record visits of ancient astronauts to Earth, a "transmission" from Timothy Leary, and much material on media, psychotechnics, and energy systems. The mix is typical of the carnival where the aim is to assemble a mental collage that packs in everything on the consciousness circuit, on the assumption that, somehow, it's all part of the "network," it all fits together.

But of course it doesn't—not as a coherent metaphysical vision, nor as a disciplined sensibility. It is simply a motley of startling colors worn by those who may vividly sense the presence of the Extraordinary in our time but who cannot yet tell the awesome from the absurd, the miraculous from the merely extreme.

A harsh judgment on what is intended by many "New Consciousness" proponents to be a significant expression of cultural renewal. But the "Oh, wow!" style of the carnival reveals a vice worse than tastelessness. For all its boasted daring, it weakens toward a small-mindedness which is finally as literal and materialistic as the most militant nineteenth-century positivism, and as great a danger to the transcendent impulse in our lives.

Let me give a few examples of what I mean drawn from some of the more prominent figures in the carnival.

ERICH VON DÄNIKEN AND HIS CHARIOTS OF THE GODS

Erich von Däniken, author of *Chariots of the Gods?*, argues, with a deal of best-selling sleight of hand, that the Earth was visited by astronauts in prehistoric times. From this event, which transpired during a primitive stage of human development, supposedly spring all our religious traditions and myths of the gods. The theory is now frequently and freely used on the consciousness circuit in much the same way that the legend of Atlantis has been used by an earlier generation of occultists—as a daring new vision of human cultural origins.[3] But it should require little insight to see that what we have here is simply an updated form of Euhemerism, that disenchanting philosophy of the ancient world which contended that all myths were based on imperfect memories of heroic human beings. Thus, religion is nothing more than bad history.

Euhemerism, in one form or another, has been invoked

3. There are now several opportunistic authors riding the von Däniken bandwagon, offering us titles like *God Drives a Flying Saucer, Gods and Spacemen of the Ancient West, The Spaceships of Ezekiel,* etc. The thesis is the same, only the exclamation points get bigger.

many times in our society since the Age of Reason, always with the same skeptical intention: to cheapen mythical awareness by depriving it of its independent and superior status as a form of knowledge. The Euhemerist spirit lives on in every attempt that has been made to subordinate myth and religion to some underlying and misperceived historical fact or social necessity. Freud was a Euhemerist in arguing that the gods were displaced father and mother figures; and so too every sociologist or anthropologist who seeks to reduce the "meaning" of religion to its structural-functional utility. Von Däniken's speculation is really no better than a science fiction version of Euhemerist thinking; its terrible literalism is a retreat from everything that Joyce and Jung, Mircea Eliade and Joseph Campbell have done to reclaim the visionary meaning of myth and to point our minds toward higher modes of knowledge. As far-fetched as von Däniken's conclusions may be, their style is that of conventional historiography; and their effect is to drive thought back toward ordinary historical analysis and explanation. But the glossy science fiction surface of von Däniken's writing blinds many of his fans to the essentially reductionist character of what he would have us believe.

Much the same criticism could be made of another leading light on the consciousness circuit: Immanuel Velikovsky, whose *Worlds in Collision* and its sequels are similar efforts to degrade mythology, in this case by turning it into disjointed reports of astronomical calamities. Velikovsky is not as sweeping in his Euhemerism as von Däniken; he takes a far smaller portion of myth into account. But by the time he finishes with the Old Testament, most of it has been turned into jumbled geology, and we are left with the distinct impression that *this* is the one respect in which myth has something real and important to tell us. The deductions Velikovsky makes are not an illegitimate use to make of myth; but his approach is nothing like understanding its real meaning. The cult that has come to surround Velikovsky in recent years is especially troubling when it takes the tack that his work has somehow ushered myth and scripture into the province of science by demonstrating what "useful" astronomical and geological documents these are. In spirit, that is not far removed from one of von Däniken's imitators, who insists that we had better begin taking the Ark of the Covenant seriously . . . it may have been an electrical generator.

WILHELM REICH AND HIS ORGONE BOX

Lately, since the publication of W. Edward Mann's *Orgone, Reich, & Eros,* Wilhelm Reich's orgone theories have come rapidly back into prominence on the consciousness circuit. Lumpen occultists have quickly enlisted orgone as part of the swelling repertory of unconventional medicines, and are once again, as in the late forties, crafting orgone boxes and orgone blankets.

The orgone theory, in its full-blown religious interpretation, was a product of Reich's later and less fortunate years, and even many loyal students of the psychiatrist—who was undeniably one of the leading therapeutic talents of our time—have been willing to admit the embarrassing eccentricity of the idea. Reich believed that orgone, an aquamarine luminosity he could see in the air all about him, is the biodynamic energy of the universe. It accounts for the creation of life and is the secret of achieving an expansively vibrant, "genital character." It is the élan vital of the cosmos, and to bask in its concentrated power, so Reich taught, will surely cure all that ails you, including cancer. It was the latter contention that led to Reich's tragic confrontation with the Pure Food and Drug authorities, by whom he was convicted of quackery and sent to prison, there to die of heart failure in 1957. By the time of his death, Reich had become passionately messianic about the new science of Orgonomy. Orgone was "the greatest discovery in centuries," and he, the first Cosmic Orgone Engineer on Earth (there were more on other planets), alone could save the world from fascism, devitalization, and global suicide. Orgonomy was not merely the energy of successful orgasm; it was world salvation. But Reich saw himself, like Jesus, suffering the crucifixion of a revolutionary prophet ahead of his time. And, indeed, there hovers about Reich in his final paranoid agony the martyrdom of a sensitive soul driven mad by the world's far greater, more destructive madness.

Now I would not myself dispute the existence of the universal life force Reich felt he had discovered. Mana, the wakan-tanka, the *anima mundi*—called by a thousand names, that vibrant energy is among the most archetypal of human experiences. It flashes through the art and poetry of every culture; it is the fire that sets alight rhapsodic utterance and ecstatic ritual. Reich only became preposterous in his mystic seizures when he claimed he had done a better job of illuminating the nature of the life energy than the innumerable artists, seers, and sages who have celebrated its awesome presence. For what was it, at last, that Reich, as engineer rather than artist, had accomplished with orgone? He had captured it in a box and taken its temperature! He had "proved" its existence by virtue of the thermometer. One recalls Goethe's wry criticism of Newton's experiments with light. Why lock yourself away in a dark room to study what fills all the heavens around us? The more so in Reich's case. Why bother with little boxes and thermometers to experience what rolls through the universe massively and magnificently? Unless, of course, one has not the capacity to experience on that scale.

Anyone who has read Reich's account of the orgone accumulator (especially the embarrassing attempt in 1941 to enlist Einstein's support for Orgonomy) cannot help but blush for the sheer silliness of the episode. It is an example of shabby science and worse religiosity. As tactfully as Einstein tried to persuade Reich that his so-called experiments were a folly, Reich could only believe that nefarious conspirators had poisoned the physicist's mind against him. But what is even more pathetic than Reich's role in the matter has been the continuing effort by his sympathizers and disciples to vindicate the unfortunate man's "research" and refurbish the orgone theory's scientific credentials. Do those who pursue the task not realize the galling disparity between the classic proclamations of the life force—the yin and yang, Blake's mythical Luvah-Orc, the landscapes of Van Gogh—and Reich's measly thermometer readings? If the Archangel Michael walked among us resplendent in his glory, would they try to verify his identity by taking his fingerprints?

Once again, as in the case of von Däniken, we have here a crude attempt to scientize and materialize a spiritual perception, with the same result of shrinking the awareness rather than expanding it. Instead of the highest and finest expression of that perception, we get a degraded, sensational substitute. And sensationalism never captures, but only caricatures the drama of a significant experience.

URI GELLER AND HIS KITCHEN SPOONS

By the time these words reach print, it is entirely possible that Uri Geller, like many another psychic celebrity, will have been exposed as a charlatan.[4] My criticism of this latest stellar attraction in the occult carnival, however, involves no judgment regarding his purported powers or those of other psychics. I take it for granted that the universe brims with mysterious oscillations and potencies our science has yet to discover—if it ever will. The psychics may well be in

4. See, for example, the devastating critique of Geller and his various investigators by Joseph Hanlon in *New Scientist* (London), October 17, 1974.

touch with forces of nature that seem as much the stuff of make-believe to us as electricity, nuclear energy, and laser beams would have seemed to our ancestors of two centuries ago—or as acupuncture did to our physicians only ten years ago. Moreover, we have barely scratched the surface of the mind and its busy interaction with nature. Why doubt that there are countless technologies of human consciousness waiting to be illuminated in the shadows of our ignorance?

Yet how sad it is, if Geller is in fact among the authentic psychics, that we must once again see these potentialities of mind reduced to their lowest, commercial terms and treated as show business amazements indistinguishable from the stage magician's stock in trade. Spoon bending, watch stopping, ring and bracelet disappearances—how much energy and attention must parapsychology continue to sacrifice to such trivial tricks and small time?

But far worse, Geller and his impresario-manager Andrija Puharich (an ESP researcher) have seen fit to build atop this small repertory of vaudeville stunts a pseudoreligious superstructure of crushing presumption.[5] Their story is that Geller has become the unique mortal instrument of the "cosmic mind" which, it turns out, is the agency behind the UFOs. (So the general law is confirmed that just as flies are drawn to dung, so all lumpen occultists of the twentieth century are drawn to flying saucers.) In Geller's account, the saucer folk are really galactic intelligences programed into an omniscient computer that has been sent back to us from the future and that has been running human history for millennia. Which means that the computer, called "Spectra," is none other than God, the very God who appeared to Abraham in Hebron. Hence, Spectra's special fondness for Israel, still the land of the chosen people. Uri Geller, promising young discotheque magician, emerges as

5. I am working here from Puharich's book *Uri: A Journal of the Mystery of Uri Geller* (New York: Doubleday Anchor Press, 1974).

nothing less than a prophet of Israel, the sole instrument Spectra will use for the next fifty years. (Such is our fate.) Upon learning this dread truth, Geller's words are reported to have been: "Believe me, this is very sacred."

And what is it the almighty computer-God does with his omnipotence during this new, science fiction dispensation—besides, that is, defending Israel against its enemies? He makes the tape cassettes, cameras, and ashtrays in Uri Geller's life vanish and reappear at the rate of about two dozen a week; he sends up flares in the desert which nobody but Geller and Puharich can see; he teleports housekeys around the globe; he clamps Swiss watches on Geller's wrist and whisks them off; at Geller's request, he repairs Werner von Braun's pocket calculator; and in one crowning miracle, he materializes a Brooklyn-made, electric massage machine in Geller's Tel Aviv hotel room.

This, bear in mind, is the veritable God of Abraham, Isaac, and Jacob, the Lord of Hosts who, in times past, revealed himself to his people by way of rhapsodic prophecy and the tables of the law, the psalms and the Song of Songs. Now, using Puharich's tape recorder as his medium (the cassettes invariably vanish after Puharich has heard them) his eloquence has deteriorated to a rather nonscriptural level; so we have pronouncements like: "Do a movie on Uri. Melanie is the one to do it. Work at it; it will come out at the right time." Or, becoming more deeply revelatory:

The ultimate powers, whether on the particle level or the cosmic level, are in rotation and drawing off of the gravitation power from the center of the system. . . . This rotation energy can be used from outside the galaxy. The computerized beings are under the direction of the "controller," or what earth man calls God, or gods. In the future this general idea would be formulated in rigorous mathematical language.

Spectra actually provides us with an example of "rigorous mathematical language." It looks like this:

$$Z = X.o$$
Where Z is the superconscious,
X is the sense.

Yet, for all his latter-day literary mediocrity, Spectra is still a God of the Book: Puharich's book, written by divine command against a publisher's advance that is doubtless somewhat in excess of anything Amos, Isaiah, or Ezekiel earned for their labors. *And* with a "Knowledge Book," which will reveal all the secrets of the saucer folk, promised for future publication—as soon as Puharich "finds" it where God has hidden it.

Surely, if schlockmeisters like Puharich and Geller wish to compound pomposity with blasphemy, they owe us some minimal awareness of the stature of that which they would defile. What we get from them instead is the lumpen-occult version of Abbott and Costello meet Jehovah: desecration at the hands of imbeciles. Again, as so often in the carnival, we are confronted with a façade of sensational claims which covers an infinite poverty of spiritual imagination. What is it, after all, that Geller appeals to but the same obsession with material power which has been the curse of our culture since the advent of the steam engine? Now the power has (supposedly) become psychic; but it draws on the same lethal fascination. If Uri Geller possessed the telekinetic power to turn Mount Sinai upside down, he would remain a mind-shrinking distraction on the Aquarian frontier. For there is less to stretch and nourish the visionary energies in his entire bag of tricks than we can find in one verse of the prophets. The total force of his presence does nothing but diminish our conception of the sacred. He is, indeed, the perfect example—perhaps the object lesson of recent times—of how psychic power (if he possesses any such) can be the annihilating antithesis of spiritual growth. As psychics, Geller and Puharich behave like petty opportunists;

as persons, they are fools who have lost all sense of truth and dignity in their pursuit of public acclaim. Those who would admit them any place in the religious regeneration of our time are bringing knaves to the temple.[6]

Demon Deus Inversus Est

I have limited myself to three prominent examples of carnival banalities, but one could go on to compile a list as long as a book of similar inanities that seek to crowd their way into the Aquarian frontier. In some cases, we are dealing with the shoddy inventions of untutored minds that have been carried away by a passion for charismatic self-advertisement; what culture has not been plagued by its share of numbskull messiahs skimming mystic clichés off the top of their own delusions of grandeur? In other cases, we are dealing with plain fraud perpetrated for cash or glory. One sure sign of both distortions is the witless mingling of high pretension and sheer trivia—as with the advocates of "Pyramidology" who intimate access to the ageless wisdom of Toth and Cheops . . . and in the next breath are telling us how we may use "pyramid power" in our own home to keep our razor blades sharp forever. Another generally reliable danger signal is contempt for all traditional teaching, a vice that usually assumes some manner of scientistic authority. The result is an effort to replace myth, mystery, and all inherited forms of spiritual initiation

6. One final and ominous fact regarding the direction and intention of parapsychic research in some scientific circles. Much of Uri Geller's reputation as a proven psychic rests on the study made by Russell Targ and Harold Puthoff at Stanford Research Institute and published in *Nature* (October 18, 1974). A representative of the Defense Advanced Research Projects Agency was invited to these experiments by Targ and Puthoff, who had every hope of securing Defense Department funding for future research. The DoD observer (accompanied by a sleight of hand magician) was wholly unimpressed by what he saw. (See letter from George Lawrence, *New Scientist*, London, March 6, 1975, p. 595.) Let the believers be warned. We have the heartbreaking example of the atomic physicists before us. What more do we need?

by research, laboratory experiments, or spectacular para-psychic demonstrations. The few remaining roots that tie our visionary energies to spiritual tradition are tenuous enough; those who would shred them still further deserve to be held suspect.

As a phase in our spiritual renewal, a bit of fun and games, some amateurish groping and vulgar improvisation, may be inevitable. There is so much we have suddenly discovered, a buried treasure of lore and arcane tradition, a wealth of laboratory findings and therapeutic possibilities. We are opening the Aquarian frontier at a mile a minute—as if we were, with feverish haste, preparing for a massive cultural transformation, trying everything in sight that might serve to guide us through the uncertain passage ahead. But it would be folly to pretend that we are in no danger of trivializing the project we are in by diluting the golden elixir with soda pop substitutes. As anything more serious than a party game, it just will not do to mix the Buddha with Buck Rogers, high mysticism with hijinks. In this area of life, as in every aspect of the life of the mind, there is an exacting standard of excellence. And excellence is not to be gained without disciplined judgment, no more than it will be won by the pedantry of unengaged, conventional scholarship.

Already we can see about us one unfortunate result that comes of falling upon too much too soon on the consciousness circuit. The several mass religious movements that have lately sprung up—especially the Jesus Freaks, the Divine Light Mission, the Krishna Consciousness Society—are clearly born of anguished reaction to the confusion and profusion of strange, new experiences. To a very great extent, these movements are functioning as drug rehabilitation programs for their youthful members: the shelter of an orderly life after the chaos of too many bad, bewildering trips. They are last-ditch sanity savers and, to that degree, a welcome mercy to their adherents. But extremes of anxiety have never been a sound motivation for choosing a spiritual career, and the true guru, while knowing how to minister to desperation, never uses its psychic leverage to gain disciples. For it is desperation that leads to dogmatism and exclusiveness, the security blankets of the spiritually infantile.

Unfortunately, sectarianism is the major characteristic of these new religious movements. The Jesus People revert to the supposed uniqueness of the Christian revelation; the disciples of Maharaj Ji insist that they alone have seen the Lord of the Universe in our time, choosing for that role a young man who has yet to do or say one thing that reveals greatness of soul; the Hare Krishnaites wield the Bhagavad Gita with all the small-minded literalism and parochial arrogance of any hard shell Baptist preacher quoting scripture. Fanaticism might seem an ironic response to the carnival's permissiveness, but it is a standard vice of religious psychology. In the midst of the deluge, one can so easily mistake the first passing timber for an entire continent.

One final danger of the carnival. Every religious tradition warns us at some point of the perils of demonic interception. We do not wander into these strange regions without risking mind and spirit both. Lucifer, so tradition teaches us, was once a prince of light. Can we safely assume he has forgotten how to set us false beacons or to dazzle us with many a pyrotechnic display? If, in salvaging the outlawed spiritual teachings of human culture, we are to take their wisdom seriously, then we can hardly ignore the symbol of the great trickster who is also a god—but god upside down. *Demon deus inversus est.* And where else might we expect the trickster to lay his best traps than at the carnival funhouse, where everybody comes just for the fun of it, expecting a safe way out of every blind turning?

3

Freaks: The Evolutionary Image of Human Potentiality

> . . . And you're surrounded by this terrifying
> freak show
> by this terrifying freak show
> and they're ridiculing you
> and they're ridiculing you
> and they're ridiculing you.
> —John Giorno, "Subduing Demons in America"

At the close of the Stanley Kubrick-Arthur Clarke film 2001, a saucer-eyed star-baby begins its mystic voyage to Earth from Jupiter to launch the next cycle of human development . . . to the accompaniment of Richard Strauss' Nietzschean superman motif. The theme of the film echoes the evolutionary fantasies of Olaf Stapleton in his novels of the 1930s (The Last and First Men, Star Maker, Odd John). *After falling out of print for two decades, now Stapleton's offbeat and awkward forays into the romance of higher evolution have found a wide new audience, new editions, and a host of imitators.*

Colin Wilson, in his recent popular study The Occult, *hypothesizes a "Faculty X," an innate superconsciousness which is latent in all people and emergent in the great psychics and mystics. Wilson calls it "the key not only to so-called occult experience, but to the whole future evolution of the human race. . . . Man's future lies in the evolution of Faculty X."*

Teilhard de Chardin, the maverick Catholic theologian

whose work draws more interest by the year, speculates that "we have entered an entirely new field of evolution" which will climax in the creation of a world-embracing "noosphere," a sort of apocalypse of expanded consciousness. His prediction is that the human race will then be collectivized into a single spiritual entity at the "Omega Point" of history, where all who accept their evolutionary destiny will experience "an ecstasy transcending the dimensions and framework of the visible universe," and the mind will be liberated from its "material matrix" forever.

Joel Kramer, a prominent yoga instructor, describes disciplined meditation in his book, The Passionate Mind, *as the key to a "new evolutionary mechanism that is based upon consciousness" and which might now evolve us away from violence and materialism. So too, Gopi Krishna, the Indian yogi, believes that kundalini is a psychophysiological energy which has carried forward the entire course of evolution. He proposes that this energy, which is the "arbiter of human destiny," should now be tapped by laboratory research for the deliberate "cultivation of genius."*

Timothy Leary, sending "galactic teletypes" from his California prison, conjectures that human development runs through seven ascending "circuits" of consciousness. At the highest "neurogenetic circuit," he anticipates the power to decipher experientially (from within) the human evolutionary future that is stored in the DNA code. (The ultimate message, as Leary reads it, seems to be that we must escape from our doomed planet and strike out via spaceship for worlds unknown.) "The first goal of a neurogenetically alert person," Leary tells us, "is to adopt the evolutionary perspective: to see the goal of mankind as mutational, to see the human role as student and assistant in the evolutionary process."

At Esalen Institute, Michael Murphy and George Leonard offer a master seminar on "The Evolution of Consciousness," a survey of "cosmic and evolutionary perspectives on the world's unfoldment." "This seminar," they explain, "will explore the thesis that a transformation of human consciousness as momentous as the emergence of civilization is under way."

Wherever the discussion of human potentialities arises these days, the evolutionary image is never far off. It emerges, as this brief sampling suggests, at a variety of intellectual levels: in offbeat metaphysical speculation, in the mystical religions, in science fiction, in lumpen-occult literature. It has become an almost casual allusion among the experimental psychotherapists. It even reaches out to tinge the political life of the day with strange, new colors. In times past, the difference between revolution and evolution was all the difference between radical and conservative politics. The two conceptions of change stood opposed to each other like the Dionysian and Apollonian virtues: to one side, spasmodic rebellion; to the other, dignified and controlled transition. Now, while political radicals continue to celebrate revolution Old Style, there is also a cultural radicalism abroad which prefers to speak of evolution—of an evolution of consciousness that will unfold with the sweep and depth of revolution, that will *create* revolution as inevitably as a ship cutting the sea turns over the waters in its wake. Perhaps the Beatles' song "Revolution," with its admonitory refrain "You better change your head instead," is the most popular expression of the argument.

An evolutionary leap in consciousness: the idea has become the updated millennialist style of the Aquarian frontier, the age-old hope of the saving remnant voiced in a biological vocabulary—that the New Jerusalem shall be achieved by an evolutionary breakthrough, that the world shall be redeemed by a contagious psychic mutation hatched in the gray matter of a chosen few. Notice how readily these days members of the dropped-out and dissenting minority speak of themselves as "freaks"—and how proudly they take the name, hoping perhaps that they are among the first of the mutants, the new humanity that holds the promise of the New Age.

Do they think of themselves as a super-race? If so, then they are a superrace that stands apart from every existing race, nation, class, culture, having no territorial imperatives, needing no political scapegoats, proclaiming themselves nobody's master. Freakishness is an allegiance pledged to a lifestyle by the culturally disaffiliated—a sensibility for extraordinary and marvelous realities, if only those that can be pharmacologically induced. In the Aquarian millennium, the sign of election is a mutation of the mind, perhaps a willful reconstitution of the genetic juices.

Now I take up the evolutionary image here, choosing it out of the welter of the Aquarian frontier as a unique insight into our historical moment. In effect, I will be identifying the Aquarian frontier itself as an evolutionary transition of consciousness—or at least its early morning light at the horizon. To be sure, the idea is still in flux, still being worked over, felt out, stretched, and refined as it passes from hand to hand. Like any idea that enters the popular awareness, it is not always treated with great philosophical sensitivity. There are varieties of evolutionary thought on the scene that I find limp, mindless, or plain comic. Leary's biochemical mysticism, for example, reads to me like pretentious science fiction; even Teilhard de Chardin's rhapsodies on the Omega Point sound too much like apocalyptic cheerleading, lacking all sense of risk or tragedy. But we are dealing here with a contemporary myth still in the making, and it is above all the popular currency of the image which lends it special value. Something in the temper of the times responds to it, something that longs for new life and higher purpose. As crude or faulty as its formulations may be, the evolutionary idea *feels* right to a growing number of those who walk the Aquarian frontier. As freaks, they feel themselves swept up in a cultural disjuncture so challenging that only a new species may be able to negotiate the transition. Had they possessed our gift of self-awareness, would our amphibian cousins as they first emerged from the sea, or our Neanderthal ancestors as they wrestled with the first rough approximations of speech, have reflected upon their situation in the same way? Would they also have seen themselves as freaks proudly serving a troubling-tantalizing new identity that stretched their potentialities to the utmost?

Why "evolution" rather than (merely) "change"? The choice of words depends upon the significance one sees in the still marginal transformations of consciousness that are happening among the early settlers of the Aquarian frontier. To call their experience "change" is to see it as an ephemeral, arbitrary, and mainly individual fashion of the times. "Evolution" implies an innovation which is prophetic, ingeniously "adaptive," and apt to become species-wide: collective arrival at a new stage of being. As I will be using the concept here, it means a revealed and realized potentiality—a step that moves us closer to the destined completion of human nature.

To the biologist, of course, evolutionary movement has to mean a genetic innovation, a physically based mutation. But that is not really saying much, since the ultimate nature and action of the genes is still a thick mystery. What alters the genes—chance, intention, or destiny? Is the genetic stuff the originating impulse or only the recorded afterthought of evolutionary transformation? Is DNA perhaps a time capsule packed with coded potentialities waiting to be triggered and released into history by need or circumstance? Can an individual or collective act of will shift the genetic material into new patterns? Are the genes, at bottom, any more meaningfully "physical" than the sub-atomic rhythms of matter? Is the genetic message more like a building harmony of musical tones than a computer code? Is behavior in the genes? Is thought? Is speech . . . creativity . . . game-playing . . . visionary power? Is *mind* in the genes? Or are the genes perhaps a matrix where the mental and physical aspects of life mingle into a reality for which our science has no conceptual framework?

All these are questions that may never be answered by empirical study, as indeed are all questions about the evolution of consciousness, insofar as we mean something other and more than an alteration of the physical brain. For standard anthropology, mind and all that occupies it belong to the realm of "culture," and culture, we are told, is not inherited by the genes but by instruction. It is not the subject matter of the natural, but of the social, sciences. In P. B. Medawar's terms, culture is "exosomatic" (outside the body) as opposed to physical mutation, which is "endosomatic" (inside the body). But clearly we need some third term to do justice to what is non-physical in human beings, yet every bit as constant as their basic bodily equipment. I mean such talents as speech, toolmaking, art, worship: acquisitions which are not, like a society's peculiar taste for food or fashion, arbitrary and culture-bound. We might call these the "endopsychic" elements of human nature: potentialities inherent within the mind. What *fills* these categories may be culturally contingent; but *the categories themselves* are defining universals of the human species: irrepressible talents of mind which every human being reaches out needfully to exercise. The contents are passed along by instruction; but instruction only answers an innate appetite to learn. And because the categories are universal, they tease the mind with possibilities of synthesis and unity. Because all cultures contain forms of art, we long to know what beauty is. Because all cultures contain forms of worship, we are called to find out the impulse that underlies them all, until, as in our time along the Aquarian frontier, our religious sensibilities become planetary in scope.

How else can we imagine such talents making their way forward into history except by "freakish" acts of genius which blazed the trail and, perhaps for generations, clung to the fringes of the existing culture or protoculture? That may indeed be the prime role of individuality in our most individuated of species: to innovate possibilities later to be adopted by the race as a whole, to present the living example that awakens the shared potentiality of all. So we would have to expect any effort to fulfill that potentiality to enter the cultural repertory through the insight of individuals and marginal groups who, until the time was ripe for the innovation to spread, might very well look like monsters and prodigies of nature.

In the case of the visionary energies, we are dealing with a special talent, one which—so I will suggest in my personal myth of the Few—may be the creative impulse behind the full range of human culture: the white light which has been broken by the prism of history into a spectrum of separate colors—art, technics, science, law, on down to the profane activities of daily life. But the primordial talent itself, the undivided white light, has lingered on among its lesser derivatives, a spiritual vocation cultivated in its essential purity by those of visionary genius. And now perhaps it finally presses forward to be realized as an historical force in its own right, a transformation of consciousness which will reveal the theme we have been varying since the beginning of human culture.

It is admittedly no better than a hunch and a hope to identify our current exploration of the Aquarian frontier as the opening of a new endopsychic category comparable in stature to the invention of speech or tool using. And yet, coming as it does at an historical moment vibrant with planetary menace—the bomb, the environmental emergency, the population crisis, the spreading rigor mortis of the urban-industrial system—this introspective and visionary adventure looks like exactly the saving potentiality the times demand of us. We desperately need to outgrow the dismal and diminished human image we inherit from the past two centuries of industrialism. We need a radically altered conception of ourselves, of our primary needs, of our place in nature, of our cosmic vocation. And here it is, fitfully unfolding before us; even in its crudest, popular renditions, it is a picture of ourselves, no longer as *homo faber* or *homo economicus,* but as humanity transcendent, seeker of meaning, creator of visions.

The chapters that follow might be called a genealogy of the evolutionary idea. They begin in fantasy and finish with some very shadowy historical interpretations. The purpose of the fantasy—my metahistorical sketch of the Few—is to suggest some seedlike, original formulation of the evolutionary image which has at least a minimal connection with historical studies and anthropology. The scholarship of Mircea Eliade, Joseph Campbell, Lewis Mumford, Siegfried Giedion, Joseph Epes Brown, and Gertrude Levy stands behind my fantasy of the Few; no doubt in some degree too the images of Black Elk (the Oglala Sioux medicine man), and of our prehistoric ancestors created by William Golding in *The Inheritors.* But I am, at last, willing to let all I say along these lines be taken as an indulgence of the imagination, a few quick sparks struck in the hopeless prehistoric darkness.

From that point on, we follow the zigzag and often subterranean course of the evolutionary idea down to the present day. Having begun (possibly) as high visionary experience, the idea first puts on the costume of myth and mystic ritual, then passes into the teachings of ancient esoteric schools (Neoplatonism, Gnosticism, Kabbalism, Hermeticism), is outlawed as a Christian heresy, resurfaces in the Romantic vision of nature and personality, is harshly literalized into a scientific theory by modern biology, becomes a modern occult doctrine, and finally arrives on the Aquarian frontier as a standard of therapy and an image of spiritual liberation. If this interpretation achieves its purpose, it will leave the idea of evolving consciousness enigmatically suspended between history and myth, scientific fact and prophetic symbol—a distant fire sign of our collective destiny whose meaning lies waiting for us in the experience of self-transcendence.[1]

1. The collection of evolutionist thinkers I mention at the outset of this chapter is not meant to be exhaustive, but only illustrative. Among others who deserve attention are Gerald Heard, *The Five Ages of Man* (New York: Julian Press, 1964), P. D. Ouspensky, *The Psychology of Man's Possible Evolution* (New York: Vintage Books, 1974), and the many works of Sri Aurobindo, starting with his *The Future Evolution of Man* (Pondicherry, India: Sri Aurobindo Ashram, 1963). One might also look at Richard M. Bucke, *Cosmic Consciousness* (New York: Dutton and Co., 1969), first published in 1901, a quaint, early effort to record "the evolution of the human mind" by an American psychiatrist who did most of his thinking on the subject in the 1870s and 1880s under the influence of Walt Whitman's poetry.

4

The Few

> The Old Men studied magic in the flowers,
> And human fortunes in astronomy,
> And an omnipotence in chemistry,
> Preferring things to names, for these were men,
> Were unitarians of the united world,
> And, wheresoever their clear eye-beams fell,
> They caught the footprints of the SAME.
> Our eyes
> Are armed, but we are strangers to the stars,
> And strangers to the mystic beast and bird,
> And strangers to the plant. . . .
> —Emerson, "Blight"

The greatest discovery we have made on the Aquarian frontier is how much there is to be *re*-discovered. Far more than innovation, salvage is the major cultural project of the frontier: the reclamation of ancient insights and primeval disciplines of enlightenment, many of them recovered in shreds and fragments from dead or doomed societies. While there is much trendy talk on the consciousness circuit about all we have learned from encephalographs and Kirlian photography, orgone accumulators and split-brain research, there has also sprung up a lively respect for the primitive and traditional. The introspective quest has become for many an archaeology of human experience, a venturesome dig into the prehistoric origins of visionary awareness.

But then religion and the occult have always been backward-looking, haunted by rumors of a primordial illumination, an old gnosis transmitted from on high to ancestors more glorified and godlike than ourselves. The Dream Time Long Ago. The Golden Age. The *Saturnis Regna*. The Age

of the Yellow Ancestor. The Garden of Eden. How are we
to regard this world-circling fairy tale of privileged beginnings? As wish? Conjecture? Why not as memory, faded
and much elaborated, but essentially sound? Certainly as
far back as we go, searching for the origins of the great
teachings and archetypal symbols, the lineage points further
back still, toward formulations that grow ever simpler, yet
more subtle, and always more tantalizingly authentic. In
the paleolithic cave paintings, in the myths of oral tradition, there glows a purer, more naïvely receptive intimacy
with the sacred than anything dreamt of since in our philosophy. How else to explain the inherited religious genius
of our race except by supposing an age of original splendor?
It is surely instructive that among the most primitive of our
fellow humans we often find the most vibrant awareness of
spiritual realities, the deepest intuition of transcendent
symbols. As if simplicity of material means served to make
space for culture of the spirit.

The ancients against the moderns. The old quarrel, as
hard fought as ever. Do we hark back or race forward to
find the great secrets? Grant that we moderns have our
urgent and peculiar role to play: ours to salvage universal
themes out of the immense and congested variety we inherit, to trace, from our vantage point in history, the contours of a planetary culture. But the substance of that culture—the undying truths and enduring symbols, the great,
high flashes of visionary energy—all this (so I am convinced) is old enough to be timeless: stuff of the world's
early morning. And to study it down to its marrow bone,
even by way of definitive editions and flawless methodolo-
gies, is not to improve upon it.

After so many generations of believing that we—the white,
middle-class, masculine West—constitute the messianic vanguard of world progress, we have no more unsettling lesson

to learn than the indispensable value of ancient and sub-terranean traditions. We have very nearly steam-rollered nature and our own species to death beneath the juggernaut advance of progress; the traditional cultures are collapsing into rubble under the pressure of global modernization, and nothing anyone can do now will happen in time to save them. We have devastated whole legacies of spiritual knowl-edge to replace them with superhighways and oil derricks—beginning with the native genius of our own American con-tinent. We are long overdue to find the discipline of a suit-able humility that will clear our heads of this fierce idolatry of the future, which is, for Western society, a "dark idolatry of self." We need to be cut down to size. But not as the Third World guerrillas and national liberators would do it —by using the West's own weapons of war and decrepit ideologies against it, imitating the wrath and roar of the beast. On the Aquarian frontier, we strike a distinctive new note in our political life: a *radical traditionalism* that uses the old gnosis as sage's scale and warrior's sword against the aggression of the technocratic orthodoxies.

Once Upon a Time

Science and scholarship can take us only so far into the shadows of the past, and never far enough. One speculative step further and we are as much in the province of pure imagination as our ancestors ever were, though without their confidence in mythical insight. All the big questions beckon us back toward the Dream Time Long Ago where we work by intuition and our guiding vision of man—per-haps with a glimmer or two from the collective unconscious. How else to begin at an inaccessible beginning?

My thinking returns again and again to a personal myth, an imaginary history, or metahistory, of the human race.

My way of teasing insights and points of departure out of the unknowable. I offer the fantasy in no other spirit than that of "once upon a time."

I imagine that among the earliest humans or near-humans, still struggling with the anxious tensions of intelligence, there were a few in whom the seed of our full and destined identity found a special fertility. Feeling the thrust and stirring of this strange growth within them, these few (let us simply call them the Few) withdrew into solitude—the cave, the wilderness, the deep forest—searching for the quiet and darkness that all growing things require to begin the adventure of unfolding. Or was it possibly the midnight privacy of the dream life they sought to recreate about themselves, having first experienced their calling in the seclusion of sleep?

Perhaps for some of the Few this unaccountable act of withdrawal cost dearly, appearing as it did to rupture the precious solidarity of the intensely social human group. They had become the first outsiders, the first outlaws. Perhaps they paid for their seeming subversion with their lives, becoming the first martyrs as well. And still, even when bought at the hard price of separation, solitude by itself might not be enough; there might be the need of sacrifice to enhance the innate fertility. So the Few gave of their physical substance, inventing severe and terrible testings: fasting, asphyxiation, flagellation, self-torment—the dread repertory of ascetic disciplines that clings to all the religions of the world and whose persistence has always baffled the squeamish modern mind. Yet here was the unsparing cultivation which unfolded the seed within and revealed most vividly to the Few the peculiar identity of their kind, the discovery that has been reborn in the experience of every saint and seer: that we are *the unfinished animal*, charged with a task of self-perfection that troubles the mind with images of godhood.

An ancient teaching, yet nothing the human intellect has since learned—certainly none of our latest findings or spectacular breakthroughs—surpasses the importance of this primal truth or diminishes its awful relevance. As we are, as we discover ourselves in the world at any stage of life, we are incomplete—and so we remain until, by an act of reflection and decision, we resolve to achieve our being. Alone among all the creatures, we can fail to become what we were born to become. Not by dying before our time (our authentic maturity is never guaranteed by age) but by failing to find the depth dimension of whatever interval is ours. What comes to every other living thing as a free gift—the fullness of its identity—has been denied to us, or rather held away from us at a difficult distance . . . so that we must reach out, must *choose* to reach out strenuously, beyond where we stand. If the fulfillment of one's highest nature is the meaning of salvation, then the beginning of humanness is to know that our salvation is a task of the will.

How to envisage these mythical Few to whom I imagine the first vision of human completion was entrusted? Perhaps the shamans of surviving primitive groups are as close as we can come to their likeness. Trickster and sorcerer, prophet and yogi, all are mixed in the shamanic role. Shamans embody the crazy wisdom that strikes among humankind as unpredictably as lightning and has nothing to recommend itself but the sheer personal authority of the one who voices (or babbles or sings or dances) its word of power. So strange and solitary is the life way of shamans, so dangerously close to what social convention regards as disease or madness, that it is almost as if they were of another species: mutants who live among visionary landscapes, obedient to a different order of necessity, their eyes forever on the movements of invisible beings. Igjugarjuk, a Caribou Eskimo shaman, says of his thirty days of fasting and cold in the Arctic wastes,

My novitiate took place in the coldest winter, and I, who never got anything to warm me and must not move, was very cold, and it was so tiring having to sit without daring to lie down, that sometimes it was as if I died a little. . . . True wisdom is only to be found far away from people, out in the great solitude, and is not found in play but through suffering. Solitude and suffering open the human mind, and therefore a shaman must seek his wisdom there.[1]

". . . Sometimes it was as if I died a little." The making of shamans happens between death and resurrection. The lore and the images change from culture to culture, but traumatic rebirth is always the heart of the shamanic experience. What so many seek today in harsh therapeutic marathons echoes a rite of transcendence that belongs to our cave and forest beginnings: to die to one's old self, to put on the skin of a higher identity.

In recent years, our literature, both fact and fiction, has been remarkably preoccupied with shamanic figures. Carlos Castaneda's Don Juan is such an uncanny type, a *brujo* of the Yaqui Indians. So too Black Elk and Crazy Horse, medicine men of the Oglala Sioux. So too (more tragically) William Golding's Simon in *Lord of the Flies*, the boy who reads the world's inscrutible meanings while in trance and is lynched by his terrified comrades for revealing his fearful knowledge.

But—surprisingly—few shamans seem to finish as martyrs. Most are given generous, if somewhat wary respect by their tribes as sages and healers. So much so that their powers of influence and leadership are often immense. Why? Why does society make place for this alien breed, these psychic freaks? What is the spell their "strong medicine" casts over us, even now in this age that supposedly knows better than to take witch doctors seriously? Is it, perhaps, the spell of

1. Margot Astrov, ed., *American Indian Prose and Poetry* (New York: Capricorn Books, 1962), pp. 298, 300.

potentiality playing upon our incorrigible mammalian curiosity? What they have seen, we would also see. Where they have traveled, we would also go, despite the perils. They tell us of dream journeys to the abode of the gods beyond the sun or beneath the sea or under the Earth. They describe spectral terrors and wonders encountered along the way. We listen, and all at once the world becomes *more* than it is within our ordinary experience. *We* become more than we are in the ordinary round of life. They make us the companions of visionary beings, intimating a life beyond the flesh. All this calls out to us like a far, inviting horizon, spurring us to press our capacities to the breaking point, to reach beyond ourselves, to *grow*—and we *like* the feeling. Though we (not being among the Few) often tire and fall back from the demand, again and again we muster our resources to those high, bright moments of life when we delight in the ordeal of stretching ourselves. It is the taste of excellence we learn from the example of the Few: self-discipline born of the passion to achieve. Behind every drive for *more* and *better* stands the transcendent quest, the pursuit of the highest of all goals: to be godlike among the gods.

It is a simple enough lesson—that there is something more we must become by reaching within and then beyond ourselves. But how might that truth have been taught to those who had not been called to solitude and sacrifice? How does the pure, inward experience become an image that can be held up to others to awaken in them something of the same aspiration? If thought and speech had not yet achieved an advanced level of metaphorical richness before the Few happened upon their vision, they may have done so now. Because to image forth the visionary reality, itself beyond words, it has always been necessary to use words in a startlingly new way. Not to say what is or was, but to say "it was *as if*," thus to use speech to say what is, seemingly, beyond its powers. A leap of the mind. A poetic invention, linking the unseen with the familiar. A *symbol*.

The symbol of the heroic quest is such a transcendent use of language, a tale that gives the inner experience the shape of an external adventure—and perhaps in so doing creates the very ideal of adventure, that passion to challenge the unexplored which has finally carried our astronauts beyond the Earth itself. So too the myths of dying and resurrected saviors, of victorious combat, of dangerous voyages, of magical flight, of journeys to the underworld. All are so many symbols giving dramatic elaboration to the visionary moment. As are the art and sculpture that provide symbolic thought with a visual embodiment. As are the rituals that enact the inward exploration: the dance whose beat and frenzy induce the raptures of trance, the sacramental feast whose drunkenness or psychedelic elation mimic the ecstatic flight of the mind.

How much of our higher culture might come down to us from such a source—from rites and images originally intended by the Few as a language in which to teach the truth they had discovered? Is it possible that the entire inventory of human culture radiates out from the white-hot visionary experience of the Few, cooling and dimming as it moves further from its origin, until much (though never quite all) of the initial brilliance has been dissipated? Perhaps, then, it is still there waiting to be discovered behind the million transformations of our cultural history—the original message whose urgency first forced the mind to invent its many means of self-expression. Certainly if we look just below the surface, we can discern behind science and philosophy the solidifying forms of myth and sacred symbol, teasing thought toward structured systems of speculation. So too, behind law and jurisprudence, we see the formalized patterns of rite, ritual, and taboo. And behind the intimidating sanctity of priest and king, we see the charismatic charm of the Few hardening into political authority and social privilege.

This is, of course, an eccentric way to read the meaning of culture. Instead of starting with the structures and functions of society, to which anthropology and the behavioral sciences give paramount attention, I am reading down and away from the visionary moment. I am accepting that moment as supreme and self-contained, and interpreting the utilitarian aspects of culture as derivative, as lesser reflections of an originally spiritual experience that only later acquired "usefulness," perhaps by way of grotesque distortion. All of which is to suggest that human beings are religious before they are practical. Or rather that the religious life arises in their pristine awareness as their first, most practical concern beyond bare animal subsistence, that they find the raw materials for other uses and institutions within their religious experience. First visions, afterward tools. The mandala before the wheel, the sacred fire before the roasting pit, the worship of the stars before the calendar, the golden bough before the shepherd's crook and royal scepter.

Imagine the Few as creators of the original human culture. Imagine that culture as wholly *nonmaterial,* a subtle mindscape of myths and rituals, of cosmic wanderings and spectral explorations, of dance, chant, and rapture. Perhaps its only physical element was the shaman's body transformed into an ecstatic instrument, a magnificently articulated focus of meditation. As we see among the yoga and vajrayana masters: though they give up all material possessions and resort to absolute seclusion, nevertheless they have in their own bodies an ingeniously elaborated interior universe. This would have been a highly perishable culture, capable only of acroamatic transmission; and yet a culture possessing a sufficiently intoxicating richness to bind the fascination and the energy of our species for countless generations before the first rock was chipped into an ax. Before the paleolithic, the *paleothaumic:* the era whose only technology was magic.

Remarkable, is it not, how those who follow the transcendent pattern labor always to strip away the physical paraphernalia of life, simplifying in the direction of just such a nonmaterial culture, one that finds all it needs in the powers of the visionary mind/body? Are they perhaps finding their way back to the ascetic beginnings of human culture, where what later ages would call "self-denial" was in reality self-fulfillment? Is their path the way we must travel now if we are to end the destructive, industrial binge we have been on for the last century?

Pure hypothesis. Not even that really. Only the rough sketch of an alternative sensibility from which, perhaps, insights might flow. Regard the human animal as, essentially, a maker of tools and exploiter of nature, and one conception of life follows—in fact, the prevailing conception of industrial society. Regard the human animal as a seeker of visions, and a very different conception follows, one that calls deeply into question much that we have come to accept as realistic and rational.

Evolution in a New Key

Having come so far, let me use my metahistorical fantasy to support the weight of one further unwarranted speculation. Imagine that what emerged so long ago in the unique awareness of the Few was the next, inviting step in our evolutionary course, that from this point forward the story of human development shifted to a new key, becoming an exercise of the mind and spirit that would no longer significantly be concerned with the remodeling of bones and tissue. By which I do *not* intend the familiar notion that, within society, the influence of natural selection yields to cultural choice. That is true enough; but it is a neutral truth, one which divorces culture from nature so completely that we are left no way to assess the natural propriety of cultural choice. I mean to suggest that, with the appearance of the visionary energies, a possibility of human development opens which is *normative,* in the same way that growing into a healthy oak tree is normative for an acorn. The flowering of the visionary energies is just such an imperative potentiality; it presses from within to be unfolded, a seed of our being waiting to be enlivened in the individual awareness. Society, as it gathers around the human band generating its own necessities of survival and organization, may facilitate that flowering. But, more often than not, it hinders the growth by burdening us with inferior loyalties, unworthy distractions, urgent injustices. That is one important respect in which the visionary quest must be *transcendent;* it must strive to rise above the endless, soul-stifling emergencies of history and society. And in so doing, it provides us with that prophetical sense of the absolute by virtue of which we presume to judge history and society, claiming independence of their demands—a sense which can live on vividly even in the most antireligious forms of radicalism. How many revolutionary spirits know the debt they owe to the hermits and the desert saints of the visionary tradition—those lonely sages who first of all stood outside society and denounced its moral failures?

It is, of course, unfashionable to the point of heresy to introduce the idea of destiny into the evolutionary process. On this I will have more to say in the chapter to follow. Here, let us only observe how intriguingly the evolutionary story, as our science now reads it, describes an ascending order of mindlike receptivity, a steady increase of the power to perceive, discriminate, and make meaningful use of the world roundabout. Living things in their evolution increasingly expand their ability to tune into the universe and pick out messages that wait there to be found. At first, at the most primitive biotic levels, the communication involved may be no more than the stereospecific ability of protein molecules to "recognize" (it is the word microbiology uses, for there is no other that will serve) other molecules by their shape, or the capacity of enzymes to discriminate among metabolic reactions with absolute specificity. But there we have a rough beginning of mindlike behavior which will later build into tropisms, instincts, simple intelligence, abstract intellect, extrasensory powers. What else is the universe about us but the hierarchy of mind?

Let us borrow a technological image to express what seems to be unfolding in this process of ever-expanding receptivity. It is as if the mind were a radio receiving apparatus being assembled in an environment already filled with broadcast signals. Millions of messages fill the air available for tuning in. But, to begin with, the primitive apparatus can receive only a few, crudely empirical signals —signals based on shape or odor, chemical compatibility, light and dark sensitivity. Gradually, as the receiver becomes more complex and powerful, it draws in more refined broadcast bands and elusive wavelengths, until at last it has drawn in a vast, subtle realm of linguistic and numerical symbols: "pure ideas" fetched like magic out of the thin air of the mind's own cogitation. At each stage along the way, the receiver, pressing forward into the perceptual margins of its powers, finds its way to meanings, whether crude or refined, that were "there" in a dimension of existence that could not come through until the mind was ready to tune them in. This, perhaps, is the sense behind Plato's realm of Ideas: the feeling that the human mind, under persistent pressure of the dialectical process, grows into ever more subtle noetic experiences, and at last into ecstatic insights that must always have been there waiting to be discovered.

The Few would then be those whose awareness was first attuned to a range of knowledge that transcends the empirical, the linguistic, and the quantitative. That act of transcendence might express itself in different forms at different times and places, needing to work itself into the cultural context of those who experience the call and of those whom they instruct; the knowledge must be translated to be communicated. But the translations are from a common text which universally teaches the incompletion of the human animal in time and society, the need of the person to achieve what death cannot undo. It is remarkable how the themes of the transcendent vocation repeat across time and space. The startling summons from on high . . . the retreat into solitude . . . the ascetic regimen . . . the demonic temptation . . . the moment of divine communion . . . the assumption of charismatic authority among those who attend and hear . . . and always, always the teachings that demand an etherealization of life. Lao Tzu, the Buddha, Jesus, Milarepa, St. Anthony, Mohammed, Black Elk, Ramakrishna . . . and how many anonymous others before the dawn of recorded history? Why have we (beyond the work of a few mavericks like Mircea Éliade and Carlos Castaneda) no anthropology of this remarkable human constant—an anthropology which recognizes the visionary genius as the first world citizen, the prophet of a planetary culture we have yet to see born?

The past two centuries have offered us many keys to the human personality. We have been told that we are "naked apes" and "meat machines," creatures "beyond freedom and dignity" governed by the reflex arc or neural feedback, by class conflict or sexual need, by economic self-interest or biological competition, by cultural conditioning or historical necessity. Curious how so many modern images of human nature seem determined to root out our deep intuition of freedom and higher purpose. None of these images explains the shaman's passion for solitude and sacrifice, except as neurotic deviation; none of them explains the abiding human fascination with this universal figure. And so none of them does justice to our yearning for completion at a higher level of being. None waters the seed within us that needs to grow.

Perhaps, if repeated often enough with sufficient authority, such nihilistic nonsense will finally have its way with us, will finally scale us down to the meager and absurd thing it insists we are—the blighted identity on which so much modern literature has been raised. Kafka, Beckett, Genet, Pinter. . . . And yet, it is striking how durably tough the visionary image of human nature is and how well it has stood up to its discreditors. Even when that image loses its popular familiarity, it survives like a spore through hard seasons of skepticism and disrepute. Notice how once again it appears among us now, blossoming in paperback editions along the fringes of our dark and despairing society. Sufism, Tantra, yoga, Zen, the old gurus and occult masters—they are all back with us, a hundred years after Marx decided they were no better than an opiate, fifty years after Freud predicted that these "illusions" had no future.

To what conclusion does my metahistorical fantasy lead me? It leads me to choosing the vocation of the visionary Few as the touchstone of our true identity. What are we most essentially? Before all else, we are *meaning-seeking* creatures. As fiercely as our flesh needs bread, our activity in the world needs a justifying purpose that is not arbitrary or contingent or bound by mortality. And there can come crises when the very act of drawing one more breath depends upon finding that purpose.

If something like what I suggest about the role of the

Few as culture makers were true, then that purpose lies hidden all about us in our culture like the original text of a palimpsest, buried and written over a thousand times. But of the cultural dynamics of visionary insight I will say more in a later chapter.

5

"Try Not To Forget"

> There are people in the world all the time who know. . . .
> But they keep quiet. They just move about quietly, saving
> the people who know they are in the trap. And then, for the
> ones who have got out, it's like coming around from chloro-
> form. They realize that all their lives they've been asleep
> and dreaming. And then it's their turn to learn the rules
> and the timing. And they become the ones who live quietly
> in the world, just as human beings might if there were only
> a few human beings on a planet that had monkeys on it for
> inhabitants, but the monkeys had the possibility of learning
> to think like human beings.
> —Doris Lessing, *Briefing for a Descent Into Hell*

Destiny, Value, Intention

To speak of an "evolution of consciousness" would seem,
inevitably, to tie our thinking to Darwin and to modern
biology in general. With that connection may come a re-
assuring feeling of scientific security—until we look more
closely. For, in fact, established evolutionary thought will
not gracefully contain the idea of evolving consciousness,
unless we agree to trivialize the discussion by defining "consciousness" as the selective advantage of superior intelligence. But if we do that, then the mind is reduced to a sort
of tool, a computer adjunct of *homo faber*. Not only does
this prejudge how the mind was initially used (for its
original task may after all have been art, play, song, reverie,
meditation), but the hypothesis leaves us with no way to
explain why any creature more intelligent (let alone more
aesthetic or philosphical) than Neanderthal man should
ever have appeared in the evolutionary repertory to open

up so astonishing a gap between ourselves and the chimpanzee. True, there is a principle in anthropology called "hypertrophy"—overdevelopment—which might be invoked to cover the rapid elaboration of human brainpower beyond anything survival demands; but, in truth, to name the phenomenon is in no sense to explain why, within the terms of selective advantage and adaptation, nature takes such obvious delight in excess.

If, on the other hand, we try to press any more ambitious notion of consciousness than functional intelligence upon orthodox biology, the idea acts like a germ upon the bloodstream: it is recognized at once as a threatening intruder, and an army of intellectual antibodies swarms to devour it. If we do not admit this fundamental antagonism—an antagonism that forces us to challenge the sufficiency of orthodox biology and, if possible, transform its sensibility radically—then we risk turning the concept of evolving consciousness into a banality. The evolutionary image is a fashionable costume for philosophical thought to put on; but as long as the image is no better than a biologist's borrowed clothes, it will always be suspiciously ill-fitting, if not downright absurd.

When we speak of an evolution of consciousness in the most ambitious sense of the term—as the unfolding of our philosophical, aesthetic, and visionary talents as well as our tool-making intelligence—we cannot help but call upon three principles that have no place in ordinary biological thought. Destiny, value, intention. Darwin's role in the history of science was—precisely—to beat these three shady characters from the biological household as if they were a gang of beggars at the doorstep. His goal, and that of orthodox biology ever since, was to portray evolution as exclusively a matter of chance and undirected selection, a mindless, value-neutral process based on the random appearance and preservation of favorable physical attributes. At its most sophisticated, neo-Darwinist biology dissolves organisms and species into the abstract dynamics of gene pools—in much the same way that physics dissolves gross physicality into the ghost dance of the quanta. At this rarified level of biological discourse, the crude but dramatic metaphors of classic Darwinism ("the struggle for existence," "the survival of the fittest," all talk of "higher" and "lower" organisms) vanish into the statistics of population genetics, freed once and for all of moral connotation.

Since the emergence of Darwin's work, modern biology has been involved in one long running battle against those who would in the slightest degree moralize evolution, and so deprive it of its scientific objectivity. The crassest attempt to do so was that of the Social Darwinists, who sought to turn natural selection into a cosmic endorsement of capitalist competition and imperialist exploitation. But this was never more than yellow journalism afflicted with delusions of philosophical grandeur. More interesting by far were the efforts of Nietzsche and of Henri Bergson to bring ethical value and a sense of the grand design into the evolutionary drama. Both philosophers (along with George Bernard Shaw and the Vitalist school following Bergson) built essentially on Lamarck's theory of evolution, Darwin's main opposition within biology during much of the nineteenth century. Lamarck had placed innovative intention at the center of his thinking about variation, contending that plant and animal forms draw upon a *sentiment intérieur* to evolve the organs or capacities they need, and then bequeath their "acquired characteristics" to their progeny. Rhapsodizing freely upon Lamarck, Nietzsche insisted that evolution pointed beyond physical adaptation toward ever greater creative excellence, toward the superman who was a tyrant-artist and pioneer of self-realization, toward the saint and sage who were inventors of higher, more exacting values (but whose victory in the evolutionary struggle was by no means guaranteed, hence Nietzsche's shrill desperation). Bergson, in his turn, believed that, under the sway of the élan vital, evolution labored against the resistance of matter toward ever higher philosophical enlightenment, thus making the universe "a machine for the production of gods." Mix Lamarck, Nietzsche, and Bergson together, and the result is a world where living things willfully seek their own ascending genetic change in obedience to a destiny that, once glimpsed, prescribes the meaning of life on Earth. And that, from a Darwinian viewpoint, is a world that cannot be admitted to exist: an absolute heresy.

When I learned my basic Darwin in high school—and neo-Darwinism and new genetics in college—I learned it as dogmatic truth, as I might have learned a religious catechism. Not in the sense that no physical evidence was adduced—there was enough of that—but in the sense that no alternative theory of the evidence was ever introduced, no critical examination of assumptions and incongruities ever encouraged. Indeed, I was led to believe that the only alternative to orthodox biology was biblical fundamentalism and the "creationist" movement. Darwin's Victorian antagonist Bishop Wilberforce was held up as the epitome of anti-intellectualism—without any mention that even the good bishop (while no giant mind by any standards) raised his scriptural objection to Darwin only as the *last* of several criticisms. The others (apparently suggested to Wilberforce by the anatomist Richard Owen) included a number of shrewd thrusts at Darwin's theory and evidence, some of which could not be parried by Darwin's defenders then, some of which still stand as serious reservations to the present day.[1]

Only later reading acquainted me with the fact that,

1. The Bishop's review of *The Origin of Species* appears in *The Quarterly Review,* July 1860, vol. 108, pp. 225–64. It is very nearly a model book review, courteous, learned, incisive.

without reaching beyond science, one can find, within the province of professional biology itself, a rich body of speculation and research that directly challenges neo-Darwinism for its incompleteness, inconsistency, or shallowness. We have, for example, the morphic biology of Adolf Portmann, Edmund Sinnott, and Lancelot Law Whyte; the organismic biology of D'Arcy Thompson; the nomogenesis of Leo S. Berg; the aristogenesis of Henry Fairfield Osborn; the systems theory of evolution of Ludwig von Bertalanffy; the preformationist evolution of E. L. Grant Watson; the emergent evolution of C. Lloyd Morgan, J. Arthur Thompson, and William Morton Wheeler; the theory of telepathic species-blueprinting of Alister Hardy; the neo-Lamarckism of Herbert Graham Cannon.

None of these men can fairly be dismissed as a crank; all are scientists in good standing, fully respectful of scientific conventions, fully conversant with standard evolutionary theory and its evidence. They simply work from different theoretical paradigms which dissent sharply from the neo-Darwinian mainstream. In short, they are the Copernicans of modern biology, at odds with the Ptolemaic orthodoxy of their colleagues. And for that reason, their work is allowed to play no part in the standard biological curriculum. While there are differences among these dissenters, their paradigms can be said to have this much in common: they hold that evolution cannot be understood adequately without introducing some integrating force (either working from within living things or lying in some general agency or principle of nature) which acts to pattern the processes of biological change. It is something like Aristotle's *final cause* they would bring into the evolutionary picture, the principle which places the whole before the parts, the end before the means. In general, they all handle that controversial principle gingerly, seeking to give it an acceptable scientific formulation. But no matter how cleverly one rephrases the idea, no matter how impersonally one speaks of it, final cause must at last imply *intention* in nature: some force that moves in a mindlike way to achieve a purpose.

And that is why their thought is so readily rejected by mainstream biology. For if even a whisper of intentionality slips into evolutionary theory, all the great questions of destiny and value will not be far behind, and with them all the nasty philosophical conundrums which conventional science prefers to avoid. It was precisely to sidestep such questions that so many scientists rallied to Darwin's banner when it was first unfurled, even before the basic terms of his theory had been coherently defined and long before any genetic knowledge was on hand to make convincing sense of how variation and selection might occur. But then it did not really matter that, at key points, Darwin's theory boiled down to empty tautologies and unproven assumptions. It did not even matter that Darwin had no reply to offer Alfred Wallace, the co-discoverer of natural selection, when Wallace admitted that neither he nor Darwin could account for such extraordinary "overdevelopments" as the expansion of the human mind beyond anything physical survival demanded. What *did* matter was that Darwin had fashioned a doctrine of evolution that was *objective* and *secular*— meaning devoid of value and (above all) of God. What C. H. Waddington once said of the hard-nosed Darwinists of a generation ago was true from the start: their theories were so many "old bones" in which they might "deck themselves out as terrifying witch doctors who could conjure the souls out of the bodies of the innocently religious."

Still today, the standard reply of the neo-Darwinists to every objection and anomaly their critics raise is: "Given enough time, chance will account for it." All the marvels and mysteries of organic coordination come down to the simple formula: random mutation (or random variation, or random hybridization) plus lucky selection. Within recent

years, the concept of evolution has even been extended—
almost by sleight of hand—to include "chemical" or "pre-
biotic evolution," meaning the development of the giant
molecules on which life is based. Here, the basic Darwinian
concepts of natural selection and adaptation make no sense
whatever, since nonvital chemicals have before them no
necessary task of surviving and reproducing, and so are
exempt from extinction in any biological sense of the word.
On what, then, does selective pressure act in this suborganic
sphere, unless upon some inadmissible tendency (intention?) of atoms and molecules to create complex forms and
hierarchical structure in the universe? But here again we
are to believe that by pure chance the genetic chemistry
of life has simply fallen into place—just as the collected
works of Shakespeare would surely be produced by the
proverbial roomful of monkeys banging on typewriters . . .
if given enough time.

The irony is devastating. The main purpose of Darwinism
was to drive every last trace of an incredible God from
biology. But the theory replaces the old God with an even
more incredible deity—omnipotent chance. Anyone who can
believe that purely random combinations of genetic traits
account for the evolution of any organ or any species or
for the vital chemistry of the cell is surely as gullible as it
is humanly possible to be. That Bishop Wilberforce's Lord
God Jehovah pokes and prods our genes into preordained
patterns from atop Mount Sinai is as plausible a hypothesis.

What we have in Darwin's natural selection is a way of
understanding how the pressure of the environment can
play a part in sculpturing the physical features of an organism. What we do *not* have from the theory (nor from
the molecular biology which tells us how these features are
preserved and transmitted) is any understanding of how
minute genetic changes in individual organisms achieve the
form of viable new species, or how the species have managed

to compose a coherent picture of evolutionary development. The theory gives us no idea of what life on Earth is out to do with itself; no idea of why it ever became anything at all beyond comfortably adapted one-celled organisms. Why are we not, after all, a planet of nicely stabilized protozoa? Yet the most striking feature of evolution is—*evolution itself*, the thrust and risk of the process, the hazardous adventuring forth into ever more complex shapes and new life spaces which is the very essence of species making and of mind making. At most, natural selection helps explain adaptation; it does not explain evolution—which is the overall historical trajectory of adaptation.

Along all the lines of evolutionary development, the Darwinians tell the story of how living things become more and more ingeniously adapted to various environmental niches under the pressure of natural selection. And yet "somehow" from within this process of cautious adaptation there again and again arise life forms that are *less* intricately adapted to immediate circumstances, organisms that are more generalized or plastic, and so capable of a leap into novelty. "Advance of any sort," Julian Huxley tells us in speaking of the evolutionary process, "has to be achieved by rather improbable breakthroughs from one stabilized pattern to another." In fact, this "breakthrough" that allows any life form to vault the rigid species barrier seems to be very like the quantum phenomenon in atomic physics: not a smooth, continuous transition, but a buildup of energy and, at the critical point, a sudden leap into a new pattern. But what is the quantum of biological energy that catapaults life forms into novel mutations? Ernst Mayr speaks of "preadaptation" as the secret of mutation: the creative use by organisms of existing attributes to undertake new functions that will allow them to venture into new life spaces, which then favor certain genetic changes. (The idea has more Lamarck than Darwin in it.) Seen (or felt) from inside the creature's consciousness, what would such "preadaptation" be but an experience of *will* responding to a felt potentiality?

We are told by the geneticists that mutational plasticity and variability are the result of "errors" in the complex transcription of the DNA code which may, once in a million times, turn out to be advantageous. "Fortuitous perturbations," Jacques Monod calls them, or simply "noise" preserved by DNA, "that registry of chance." Is this not a remarkable fact: that the life process on Earth should have preserved this massively disadvantageous capacity for transcriptional slip-ups, this area of dangerous freedom which only rarely provides any species with some desirable mutation? Surely one would expect that natural selection would have eliminated such genetic imprecision back in the days of the original amoeba, replacing it with a simple and stable hereditary mechanism. Suppressing mutational error would almost seem to be the first order of evolutionary business— if indeed what we are dealing with here *is* "error," rather than the free play that unfolding potentiality requires. It is certainly not difficult to imagine a world where genetic invariance really was invariant, and where evolution had stopped at the level of a perfected protozoa. Of course, with the benefit of some billions of years of hindsight, knowing what such transcriptional errors have been capable of creating, we can now see that absolute invariance would hardly have made for the most interesting evolutionary story, since it would have ruled out all possibility of significant development, including the development of human beings. But once recognize the fact that adventurous development rather than complacent adaptation is the essence of evolution, and one is not far from agreeing that, in addition to negotiating with the environment outside, life on Earth has been concerned to express an inward need—the need to innovate and grow ever more complex.

And how are we ever to know what capacity this need—

this will to innovate, wherever it may faintly emerge into consciousness—may not have to shift a grain or two of the genetic stuff, possibly, as Alister Hardy has suggested, by some telepathic "pool of experience" shared by the species? [2] The mind, without knowing how it does so, can worry the body into ulcers and asthma; without knowing how it does so, it can slow the heart and cool the metabolism; under hypnosis and without knowing how it does so, it can raise blisters on the unburned flesh; under operant conditioning and without knowing how it does so, it can drive off addiction and change sexual responses. All this we have learned about the mind, and yet we so obviously know only a poor fraction of its nature and capacity. How far-fetched is it, then, to imagine that organisms—or at least a strong-willed Few of them pioneering their way toward an interesting new environment or an inviting new behavior—might not, after some generations of trial and error, at last nudge their nucleic acids into the shape of a suitable mutation? Until there might finally emerge a human animal of rare sensitivity whose curiosity could sense the existence of environments no longer physical, where the adaptation required of the species was a subtle change of consciousness.

But this is to speculate beyond anything orthodox biology can tolerate, for the biologists, like all scientists, prefer to work from paradigms that guarantee neat, nonphilosophical solutions. So let us, then, with full malice aforethought, cut ourselves free of the restraints of scientific respectability. Because it is not modern biology that has anything to tell us about the evolution of consciousness. If we would speak of that, we must turn to a much older tradition—an occult tradition which embraces such exotic, ancient schools of thought as Neoplatonism, the Kabbala, Gnosticism, Hermeticism. We have no one name for this sprawling but tightly

2. Alister C. Hardy, *The Living Stream* (London: Collins, 1965). See his chapter on "Biology and Telepathy."

interrelated body of metaphysical insight and rhapsodic speculation. Because it has always emphasized the "occult" or mysterious character of its teaching (we will inquire into the original meaning of "mystery" in a later chapter), I will call the tradition as a whole "the Hidden Wisdom." My treatment here is only of the most general themes that run through the several schools; yet it will be enough to show that it is from this shadowy source that the modern West inherits the original notion of evolution, the root concept which Darwin and his successors have abbreviated into the lesser idea of natural selection. All that the discussion of evolving consciousness wants to become, and all that ortho-dox biology forbids it to be—which is an examination of spiritual need and human destiny—finds its home in the Hidden Wisdom.

The Hidden Wisdom

At the heart of the Hidden Wisdom resides one supreme teaching: that the world is a cosmic drama of *transforma-tion*. The sacred does not confront humanity as a reservoir of fixed and frozen attributes; nor is human nature to be understood as completed and static. *Time* and *experience* make a difference to the cosmos of the Hidden Wisdom; they draw God, humankind, and nature together in a vast epic whose purpose is to radically transform each into some-thing wholly and marvelously different from what it now is. Dynamism, growth, unfoldment—these are the realities in which the Hidden Wisdom deals. The cosmos unfolds or-ganically, and there is a direction to that unfolding; it moves toward spiritual crisis and resolution. What human beings and the world about them are is not what they are meant to be. So there is a project which they have before them, a transformation they must effortfully achieve if they are to rise to a higher level of being. Or rather *return* to that level, for the tradition conceives of our destiny as the restoration of an original and lost perfection. In contrast to Darwinism, which teaches us that the human has arisen out of the subhuman, the mind out of the mindless, the living out of the inert, the Hidden Wisdom endorses the nearly unanimous view of world mythology that the human has *descended* from the superhuman, from the godlike and unblemished, by way of some primordial "fall." It regards the coming together on Earth of spirit and matter as a risky venture, a destined but difficult marriage which offers us the opportunity to reclaim paradise from the mortality of physical nature, but which has also cost us the memory of our original nature. We are like agents dispatched on a perilous mission with one urgent instruction to guide us: *"Try not to forget . . . try not to forget who you really are and what your purpose is."*

But we *do* forget, again and again and again, and so the great transformation is constantly at hazard. The experience we know as "history," including Darwin's natural history, is the troubled, Earthbound phase of this redemptive cycle, the phase during which our potentially godlike consciousness struggles to illuminate the dark and death-ridden world of physical nature we are in. It is the cosmic interval in which we strive to teach matter the reality of spirit. As for Darwin's evolutionary theory: this would be the story of our physical making—or rather of that part of it which has been resourcefully negotiated with environmental pressures since the initial immersion of spirit in matter on Earth. It is part of the latter, ascending phase of the great cosmic cycle as seen from the *outside* ("objectively") by those whose interest is exclusively in the course of physical transformations.

In a recent novel, *Briefing for a Descent into Hell,* a work which some critics have mistaken for psychiatric autobiography and others for science fiction, Doris Lessing has admirably captured the essential experience of the Hidden Wisdom. As an agent of the Gnostic Descent, her hero struggles valiantly to preserve the memory of his "briefing" —the redemptive instruction he carries into life from his preincarnated existence—against the distractions of personal affairs and adjustive psychiatry. "Don't forget," he is ordered before the Descent by his commander "Merk" (Hermes-Mercury, the divine intelligence of the Gnostic tradition). "Keep the memory of this moment, keep it steady." And the message is embedded in him as "sealed orders": a "brainprint" of mankind's future evolution beyond its present zombie state. The brainprint is triggered in him only late in life, taking the form of a schizophrenic seizure filled with wild visions and muddled reveries. But during the ordeal he wins through to the recollection that "There's something I have to teach. I have to tell people. People don't know it but it is as if they are living in a poisoned air. They are not awake. They've been knocked on the head, long ago. . . ."

But at last his stamina ebbs and he fails his mission, as have most agents of the Earth's many "previous Descents" now lost in the mists of fable and myth. His "mad" higher consciousness is literally jolted out of him by electroshock therapy, and he finishes "fully recovered" and restored to "himself" . . . a professor of classics lecturing on the Homeric epithet.

It is because we are so forgetful of our mission that the cosmic drama of transformation must be played out; it is our education in human identity. So God, or some aspect of God, undergoes an outreaching process of "emanation"—a descent or *devolution* into time and matter. In its mercy, the sacred takes on lesser (or rather more humanly accessible) and riskier identities, dares to spread the divine substance thin, like light diffusing into the darkness in its attempt to penetrate the farthest reaches of the universe. Concurrently—so some of the myths of the Hidden Wisdom suggest by drawing upon the doctrine of metempsychosis—each human being also undergoes a spiritual transformation as he/she passes through many lifetimes. Gradually, as the fallen spirit in each of us perceives more and more of the light in the world's dark corners, it arises from its "sleep" or its "drunkenness" or its "exile" (the three most common metaphors used in the tradition for our benightedness) and experiences an ascending evolution toward ever greater and clearer self-knowledge.

The initial experience of this personal evolution is one of entrapment and degradation: a "divine discontent" with the fallen human condition. Because of this sense of having to fight one's way out of drugged sleep, there are abundant expressions in the tradition that bitterly attack life, nature, and the cruel god of *this* world, the fiend who has subjugated us to the debasement of matter. For example, in the version of the myth we find in Blake's prophetic epics, this enemy god is the jealous tyrant Urizen, the Jehovah-like lawgiver and "shadow of horror": "Nobodaddy," who is also the ominous "tyger, tyger, burning bright." All the dread, horror, and absurdity we find in the modern existentialist vision of human abandonment can also be found in the Gnostic and Kabbalistic schools of the Hidden Wisdom, though always as preliminary experiences discovered on the threshold of redemption. Beyond the exile's cry of despair lies reunion with "the unknown God" and the whole noble truth of the soul's drama. For at some point, God's merciful devolution into matter is crossed by the soul's evolution as it rises on a ladder of visionary awareness. (In my personal myth, I have placed the Few at that crossing point, the first human sleepers to come awake.) In this grand process of human salvation, nature itself is finally to be redeemed and to become the perfect mirror of the original, spiritualized cosmos, taking on a splendor that is apparent to us now only in sudden, unpredictable flashes, like the play of sparks in a heap of smoldering ashes. As the Kabbalistic tradition tells the story, our project in life is to watch for these scattered sparks, these "vestiges" of the divine, to collect and meditate upon them until we find in them the lineaments of our original body of light, the glory of Adam Kadmon, the archetypal human first created by God. In the words of the Kabbalist master Isaac Luria, "there is no sphere of existence including organic and inorganic nature, that is not full of holy sparks which are mixed up with the Kelipoth [the elements of the unredeemed physical universe] and need to be separated from them and lifted up."

The style of the Hidden Wisdom is bound to seem bizarre to us; it is indeed a weird philosophical stew of myth and luxuriant metaphor, a rhapsodic language of the inner eye. Christians may of course recognize in it many motifs that have been taken over by their theology: the perfection of Eden, original sin, the fall, the unfolding of history as a redemptive drama, the incarnation of a savior deity in Jesus, the descent of the Light into darkness, the promise of resurrection and the New Jerusalem. All this is familiar enough. What we are less apt to see in the tradition is the original conception of evolving consciousness. But that is what we have in the Hidden Wisdom's drama of transformation: *evolution with all the biology left out,* evolution as the path followed by the human spirit in its struggle to mature and come home.

The key point is the tradition's insistence that neither the human nor the divine can be understood as static and completed entities. Rather, the universe as a whole must be experienced as a dynamic process. That is the seed from which all evolutionary thought blossoms. From that teaching we derive our universal conviction that to *know* anything, we must know its origin and the trajectory of its development; we must know where it comes from and where it is going.

Knowledge is born of an eye for change *and* for the continuity within change.

At the same time, while the Hidden Wisdom abounds in organic metaphors, it is wholly anti-naturalistic; the ancient schools lack any palpable sense that this process of cosmic unfoldment might be *seen* in nature itself as it surrounds us. Once that happens, once transformation is seen to be incarnated in the world, we can readily recognize the Hidden Wisdom as the parent stock from which the theory of biological evolution springs.

Which brings us to the work of the Romantic artists.

The Romantic Connection

By the late eighteenth century, along routes too subterranean to follow in their entirety, the teachings of the Hidden Wisdom had made their way into the broad mainstream of European Romanticism. In a few cases, such as Blake, Coleridge, Goethe, and the English philosopher-scholar Thomas Taylor, we find a studied familiarity with Hermetic, Gnostic, and Kabbalistic tradition. In others, the borrowings are haphazard and gropingly intuitive. Shelley's poetry is, for example, laced through with Neoplatonic ideas which almost seem like spontaneous rediscoveries. But what surfaces everywhere in the Romantic movement is a dramatically new vision of nature, one that diverges sharply from Newtonian mechanism and which all but insured that, in the century ahead, biology would replace physics as the prevailing natural philosophy.

For the Romantics, nature was organic and personalist, a live and willful presence, perhaps a divine presence. Its energies were vital and intentional, even unruly, frenzied, ominous. It expressed itself most fully in images of forceful growth, tempestuous power, and awful mystery. Storm-battered mountaintops, gnarled trees, wild seas, brooding heaths and moors, haunted forests and glens . . . this is the familiar stuff of Romantic nature. And from it the Romantics derived a keen eye for the morphological beauties of the world, for what is irregular and unfinished, effortful and striving. It was not clockwork routine they saw when they looked at the universe about them, but a world in perpetual transformation, filled with the struggle to unfold and become. Once cultivate the eye for such an organic dynamism, and the idea of evolution is not far off.

The Romantic contribution to Darwin's thought is obvious enough and has often been commented upon. Romanticism, we can now see, brought organicism to the center of scientific attention and with it a lively new awareness for evolving continuities. Goethe, a lifelong amateur in biology, took the mutability of the species in stride in his speculations. His own studies in plant and animal morphology had convinced him that the higher life forms must have emerged from the lower as part of the upward-thrusting and shape-shifting character of nature. We also know that Darwin's favorite and closely-studied poet was Wordsworth. What has been less often observed than Darwin's debt to the Romantics is the way in which Romanticism, in its turn, draws upon the Hidden Wisdom. In effect, what the Romantics did was to channel the Hidden Wisdom's cosmic drama of transformation into physical nature; they brought it down to Earth and clothed it in the living substance of plant and animal. With a pagan exuberance, they allowed all nature to be a sacred incarnation. The Hidden Wisdom had unfolded its drama wholly in the realm of myth and poetry, in the timeless Now which is before history and above history. The Romantics turned transformation into natural history. Their vision is of a physical universe still in process, striving to make itself something that it is not yet. Therefore, its beauty, its very *reality*, lies not in what it is, but in its effort to achieve a far-off destiny. And as with nature, so with mankind. The basic human beauty is self-realization, the unending Faustian quest for experience, for heightened awareness, and the perfection of the personality.

Thus, the descent of Darwinism runs like this:

The Hidden Wisdom: spiritual transformation via emanation and metempsychosis

Romantic naturalism: self-realization via organic dynamism and creative striving

Darwinism: survival of the fittest via the struggle for existence and natural selection

For the scientific mind, this genealogy of ideas represents a steady advance toward realism and sound logic. It traces the way in which a murky piece of metaphysics has been made rigorously compatible with empirical fact. But, in truth, in the process of passing from the Romantic vision of nature to Darwinism, something vital has been lost. *Consciousness* has been brutally uprooted from the idea of transformation, and with it all sense of purpose, all sense of destiny. The result is transformation without meaning: random fluctuation, directionless drift—mere change. That is what is bound to come of making the story of evolution the story of anything besides consciousness; for the history that counts is the history of consciousness. It is only in consciousness that the world—even the world of the remote, prehistoric past as we reconstruct it imaginatively—can be mirrored and given understanding. The history of the universe is conditioned by the shape of our consciousness; we know as much reality as our experience can contain. How, then, can we pretend that the story of the universe can be told with consciousness left out or added on as an afterthought?

In the tradition of the Hidden Wisdom, transformation has to do with consciousness wholly conceived of as spiritual awakeness. The myths of the tradition tell how God reaches down into matter and how mankind reaches up toward spirit. And where the two meet, the result is gnosis, knowledge of the sacred. The Romantics, by an imaginative tour de force, achieved the ability to find gnosis in the living stuff of nature about them, recapturing our childish sense of the world enchanted. But from Darwin forward along the scientific mainstream, this entire dimension of experience is cut away from our science in favor of an evolutionary process that is limited to the random variation or mutation of physical structures as they adapt to the requirements of an indifferent environment.

What we have here is a classic example of how a spiritual teaching can be objectivized and secularized by science until it is stripped of all religious overtones. Today, we see the Hidden Wisdom through a biological filter that screens out its finest colors. Yet, when we talk of the evolution of consciousness, it is the veiled image of the Hidden Wisdom there on the far side of that filter which captures our thought, persuading us that the great life process of unfoldment and development in which we participate is much more than a game of amino acid roulette. The lesson is clear: what we borrow from science in the discussion of human potentialities, we must borrow with great care, seeking always for the traditional themes which underlie its concepts. Our science (like our mathematics and technics) is one set of variations on those themes, one oblique reading of the archetypal knowledge of the human race. We use science most wisely when we listen for the old gnosis which resonates through its every concept and theory.

6

The Occult Evolutionists:
From Secret Doctrine to
Eupsychian Therapy

The Darwinian dissenters of the nineteenth century are usually portrayed as an opposition from the right—the biblical literalists, later called "fundamentalists"—and an opposition from the left—the Nietzschean and Vitalist philosophers who sought to expand the perspective of evolution to include will, consciousness, and spirit. As we have seen, this left-wing dissent overlaps a deal of unorthodox evolutionary theory within professional biology itself; but it also connects with what might be called an opposition *from below*—the subterranean opposition of the new occult movements of the late nineteenth and early twentieth centuries. Here we find thinkers who countered Darwin with conceptions of evolution so seemingly zany that mainstream science has never troubled itself to favor them with a gesture of refutation. Conventional intellect continues to withhold its attentions from such gurus of the occult as Madame Blavatsky, Rudolf Steiner, and George Gurdjieff, for these are indeed eccentric minds that lack all the usual academic courtesies. Their philosophies draw with unrestricted license upon myth, legend, personal vision, and antique lore; and each claims access to privileged sources of knowledge available only to loyal initiates and disciples—hardly a style calculated to win the favor of skeptics or positivist critics.

Yet, if we search the strange mythological extrapolations of these occult evolutionists to discover the vision they offer of human potentiality, we may, at the very least, find them among the most innovative psychologists of our time. In their work, we see the evolutionary image being used for the first time in the modern West as a new standard of human sanity —or rather, as a newly rediscovered standard, for to a degree that far surpasses the work of Nietzsche or Bergson, they remain unabashedly loyal to the schools of the Hidden Wisdom, and so link contemporary thought on human potentiality with a rich basis in tradition. In comparison, the Vitalists and Nietzscheans only skim the surface of that tradition; Blavatsky, Steiner, and Gurdjieff probe its metaphysical and therapeutic depths to find the underlying spiritual realities. As a consequence of their allegiance to these off-beat sources, their language and allusions are quaint to our ears. They speak of "etheric" and "astral bodies" where we have grown used to speaking of ids and egos, archetypes and complexes. Where we expect clinical or laboratory findings, they resort freely to astrology and number mysticism, alchemy and Kabbalistic symbols. For these reasons, their work has remained, until recent years, a marginal curiosity in Western society. Only now does a wider public begin to see that their aim was to deal in the visionary reaches of the mind, to explore that higher sanity for which the reductionist and secularized psychology of their day gave them no adequate body of references or suitable vocabulary.

It has often been observed that Freud, in bringing the study of the unconscious into the cultural mainstream, sought to work up a psychological model that depicted the mind as an emotional steam engine; accordingly, he filled his theories with metaphors suggestive of energy flows and blockages, pressures and cathexes. His points of reference are those of physiologist and mechanist, drawing on the prevailing world view of science. But the occult evolutionists conceived of the mind as (potentially) a mirror of the sacred. So for their model, they had to reach back to other and alien traditions, improvising a vocabulary of myths and spiritual allusions which would serve where scientific parlance could not. The result is one of the modern world's most faithful reassertions of the Hidden Wisdom—and a remarkable example of how stubbornly the lore of the old tradition survives among us, continuing to draw sensitive minds to its inner meaning. If the evolution of consciousness is widely studied today with any appreciation of its traditional resonance, it is thanks to these first-generation efforts to create a psychology grounded in the transcendent impulse.

Madame Blavatsky's Secret Doctrine

Helena Petrovna Blavatsky (1831–1891; HPB, as her followers called her) has had a bad press ever since she appeared on the European scene in 1875 as organizer of the Theosophical Society. A forceful, pugnacious, and gifted personality—worse still, a forceful, pugnacious, and gifted *woman*, one of the great liberated ladies of her day—she could not help but draw withering, critical fire by her every act and word, especially when she presumed to challenge the most entrenched intellectual orthodoxies of the age. Still today people who have never read a line she wrote remain adamantly convinced she was a fraud and a crank. With the result that Theosophy, one of the most adventurous and intriguing bodies of nineteenth-century thought, has long languished in the shadow of its founder's compromised reputation.

Of course, HPB did herself no favors by the little tricks and extravagant claims she too often employed in an effort to charm her adherents. "One must create an atmosphere," she once told a colleague, and like many another guru with an urgent message to impart (including Gurdjieff, whom she much resembles in her air of mystery and imposing authority, and indeed right down to her strategic use of the penetrating gaze), she was willing to encourage the ready amazement of her audiences. But then, she was a woman surviving by her wits, living in an era when gurus of the occult (especially if they were female) were a new and suspect breed in the skeptical, materialist West. Can she be too much blamed for indulging in a bit of defensive self-dramatization?

In any case, it is not HPB's controversial reputation or personal angularities that concern us here, but rather her ideas. For ultimately she stands or falls by the quality of her thinking, all arguments *ad feminam* aside. And in this regard, she is surely among the most original and perceptive minds of her time, if also one of the most uneven.

It is seldom remembered that, in the years following publication of *The Origin of Species*, HPB was the first person to aggressively argue the case for a transphysical element in evolution against the rising Darwinian consensus. Her exhaustingly verbose style and the eccentricity of her research have cost her dearly over the years, rendering her nearly inaccessible to contemporary readers. Yet, buried in the sprawling bulk of her two major works (*Isis Unveiled*, 1877, and *The Secret Doctrine*, 1888) there lies, in rudimentary form, the first philosophy of psychic and spiritual evolution to appear in the modern West. Her effort, unlike that of the Christian fundamentalists, was not to reject Darwin's work, but to insist that it had, by its focus on the purely physical, wholly omitted the mental, creative, and visionary life of the human race; in short, it omitted *consciousness*, whose development followed a very different evolutionary path. Darwin simply did not go far enough; his was not a big enough theory to contain human nature in the round. As

HPB put it: "Darwin's starting point is placed in front of an open door. We are at liberty with him to either remain within, or cross the threshold, beyond which lies the limitless and the incomprehensible."

Against Darwin's restricted notion of evolution, HPB posed her "secret doctrine," a composite rendering of the several schools of the Hidden Wisdom (Gnosticism, Kabbalism, Hermeticism, Neoplatonism) freely and often awkwardly blended with the Oriental religions whose literature was only beginning to receive scholarly respect in HPB's day. In these ancient schools she shrewdly discerned the original form of the evolutionary image as the redemptive journey of spirit through the realms of matter. Nearly single-handedly she rescued these long-scorned and neglected traditions from the dark fringes of our culture and sought to make them living philosophy; the project was not her least contribution to contemporary thought. Unhappily, she flawed her effort by mixing the materials of her research (imperfect as they already were) with a sensational appeal to clandestine sources totally unavailable to her readers. It was a fatal mistake in an age which valued scholarly precision and "scientific" historiography above all else.

She claimed, for example, to have studied at secret schools in Tibet and to have gained admission to hidden archives which "occult fraternities" had assembled in caves and subterranean cities—collections supposedly larger than that of the British Museum and in which the records and wisdom of lost civilizations survived. Worst of all, she was not above the discourtesy of concealing the key texts on which she based much of *Isis Unveiled* and the whole of *The Secret Doctrine*. In *Isis Unveiled*, she claims to be working from a "very old book" (otherwise unidentified) supposedly written more than 4,900 years before her time; in *The Secret Doctrine*, she draws upon a text called *The Secret Book of Dzyan*, which she purports to have translated from a language known to no living scholar. She makes these astonishing claims with no explanation as to why not even a single page of the original texts could be publicly displayed—even now when she had undertaken to translate and reveal the entire meaning of these hoary works. In so scholarly an era, HPB knew that by such high-handedness she was asking for trouble; she anticipated that her books would be regarded as "romance of the wildest kind." "For who," she asks, "has ever even heard of the book of Dzyan?" Nonetheless, she forged ahead with her project, asking that her ideas be regarded as "working hypotheses" for a philosophy of spiritual evolution.[1]

The Darwinians, HPB contended, begin at the "midpoint" of the total evolutionary progression. Lacking a spiritual dimension to their thought, their approach can only treat the later, biological phases of our physical development. But even the full meaning of this phase cannot be grasped until it is paralleled by the cosmic transformations of spirit that preceded it and continue to influence it. For matter exists, in HPB's system, only to be the receptacle of spirit; it responds to the unfolding needs of spirit as part of the grand redemptive cycle. "Our physical planet," as she puts it, "is but the handmaiden of the spirit, its master." This, the Hidden Wisdom's traditional conception of evolution, stands in HPB's work as "the secret doctrine," the "primeval revelation" which she was convinced lay at the core of all religions and philosophies.

1. "Who ever even heard of the book of Dzyan?" Apparently the leading Kabbalist scholar of our time, Gershom Scholem, has. In his classic work *Major Trends in Jewish Mysticism* (New York: Schocken Books, 1961), p. 398, he identifies the book from HPB's quotations as a loose adaptation of a "pompous" Zoharic writing called *Sifra Di-Tseniutha*. This, he believes on good grounds, HPB acquired in a Latin translation from a seventeenth-century Kabbalistic source she mentions knowing. If Scholem is correct, HPB's secrecy about the text is even more exasperating—unless, once again, we accept this as an effort to create "an atmosphere," in this case a smoke screen.

. . . Our "ignorant" ancestors traced the law of evolution throughout the whole universe. . . . From the universal ether to the incarnate human spirit, they traced one interrupted series of entities. These evolutions were from the world of spirit into the world of gross matter; and through that back again to the source of all things. The "descent of species" was to them a descent from the spirit, the primal source of all, to the "degradation of matter." [2]

Yet, though this immersion of spirit in matter is, from one point of view, a "degradation," it has—so HPB goes on to tell us—been decreed by mankind's spiritual guides (the "Dhyanis" or "Mahatmas") for the purpose of vastly enriching our consciousness. By our collective evolutionary course, and by innumerable personal incarnations, we make our way through all the realms of being: mineral, plant, animal, human, divine. And it is by virtue of this hard-won "harvest of experience" that each human being becomes a microcosm of the universe. As with Pico della Mirandola's conception of a chameleon-natured humanity, our goal is to make human consciousness the compendium of all possible forms of existence. We are, in this way, the agents who elevate plant and animal life, inert and mindless matter, to self-awareness. It is exactly to achieve this cosmic heightening of consciousness that spirit sacrifices itself by its initial descent. In the words of the Kabbalistic formula several times quoted by HPB: "a stone becomes a plant; a plant, a beast; a beast, a man; a man, a spirit; and the spirit, a God."

HPB located our era at the pivot point of this progession, where spirit, having reached the human phase of its journey, is ripe to recapture the memory of its origins and to gain the leverage necessary to raise itself and physical nature to the level of divinity. This task would still take thousands of years of purification and enlightenment in HPB's cosmolog-

2. *Isis Unveiled* (Pasadena, California: Theosophical University Press, 1972, 2 vols.) I, 285.

ical system; but its end was clear in her view: a Mahayana vision of universal salvation. It is one of her best moments, if a little too breathless. The great drama continues, she tells us,

. . . until the minutest particle of matter on earth shall have out-lived its days, until every grain of dust has, by gradual transfor-mation through evolution, become a constituent part of a "living soul," and until the latter shall reascend the cyclic arc and finally stand . . . at the foot of the upper step of the spiritual worlds, as at the first hour of its emanation. Beyond that lies the great "DEEP"—A MYSTERY! [3]

It would be impossible to do justice here to HPB's entire cosmology and psychology as contained in the 2,500 busy pages of *Isis Unveiled* and *The Secret Doctrine*, as well as in the thousands of pages more she wrote for Theosophical jour-nals and her collection of "Mahatma Letters." The panorama is too vast, the insights and the oddities too numerous for comment. There is also an unresolved ambiguity that runs through her work which makes sensible summary difficult. In what is essentially a mythical armature for supporting a godlike image of human nature, HPB too often weakens toward an unfortunate literalism, interpreting her sources not as rhapsodic declamations but as veritable historical documents—and that with no little pugnacity toward every anthropologist, astronomer, geologist, and biologist in sight. For example, she tells us that four "root races" have pre-ceded our own in the evolutionary cycle, all of them prema-ture and flawed (if not monstrous) efforts to achieve a viable balance of the spiritual and physical on Earth. There are evocative aspects of this teaching, especially the sexual progression of the races from the sexlessness of the first race through the androgyny and finally the full male/female separation of its successors: in brief, a recapitulation of the

3. *Isis Unveiled,* II, 420.

sexual evolution of the species and of the human embryo which, again, contributes to the microcosmic status of human consciousness.

But HPB presents her root races as historical populations that can be located in time and place. (In succession, they inhabited the Hyperborean regions, Lemuria, and Atlantis.) Moreover, she was convinced that they had left behind surviving, subhuman remnants. Thus, the Australian and African aboriginals are such "narrow-brained" leftovers, far inferior by nature to the "Aryans" who spearhead mankind's evolutionary course. The point is a relatively marginal one in her work, but its conventional racist assumptions are nonetheless painful. Similarly, it is only in passing that she states her conviction that the gorillas and chimpanzees (as well as an obscure people she calls "Tasmanians") are the lingering offspring of illicit mating between Atlantean degenerates and "huge she-animals." But why did such a bizarre piece of speculation even seem worth the trouble of mentioning at all in a work meant to embody the wisdom of the ages? By and large, HPB had a sophisticated eye for the deep meaning of myth and symbol; but her treatment of the root races amounts to a kind of "occult fundamentalism" and is a sad example of how stubbornly even daring minds cling to literal and historical interpretations. Down to the flying saucer watchers and "mysteries of the great pyramid" cults of our day, the vice continues to plague occult groups. The need to reify myth is deeply rooted in the Western mind.[4]

4. On the other hand, it is one of the charming features of the younger occult movements, whose members often come out of highly political backgrounds, that the casually racist overtones of HPB's generation have been sharply denounced. For example, an impassioned editorial in the new occult journal *Gnostica* (February, 1975) defending paganism and the "Third World Occult Tradition," rejects everything HPB and others have had to say about root races, along with the general "cultural arrogance" of past European esoteric groups. "Racism is racism," the writer insists, even if "one is quoting the Ascended Masters or the Atlantean Adepts." In general, the new occultism is laced through with healthy influences absorbed from third world, native peoples, and women's liberation politics.

Still, with all criticisms weighed up against her, HPB stands forth as a seminal talent of our time. Given the rudimentary condition of her sources, her basic intuition for the teachings of the ancient occult schools was remarkably astute. And there is no denying her precocity in recognizing how essential a contribution those schools, together with comparative mythology and the Eastern religions, had to make to the discussion of evolving consciousness. As a salvage operation alone, her books deserve a place among those marginal modern classics we occasionally revisit to take our bearings. Autodidact and amateur that she was, she could be agonizingly indiscriminate in her research; but she was fearless, if not outrightly belligerent, in bringing the evidence of outcast traditions into the discussion of human potentialities.

Above all, she is among the modern world's trailblazing psychologists of the visionary mind. At the same historical moment that Freud, Pavlov, and James had begun to formulate the secularized and materialist theory of mind that has so far dominated modern Western thought, HPB and her fellow Theosophists were rescuing from occult tradition and exotic religion a forgotten psychology of the superconscious and the extrasensory. If one can appreciate her ungainly metaphysical speculation on no other basis, it should be seen at least as a groundbreaking psychomythology of the transcendent personality.[5]

For all her failings—and they are many and obvious—HPB dared to work on a scale that did justice to her subject. Her canvas was gargantuan, and upon it she roughed in a range of materials—myth, ritual, primitive religions, mystical litera-

5. For a compact summary of theosophical psychology (vintage 1904) see the lecture series *Theosophy and the New Psychology* (London, Theosophical Publishing Society) given by HPB's successor, Annie Besant. If we allow for the early date and discount the somewhat quaint style, it is as fresh and ambitious a treatise on the higher sanity as anything produced by the latest consciousness research.

ture, esoteric tradition, Oriental philosophy, and even a great deal of offbeat Western science drawn from the dissenting edges of psychology and biology—which makes her among the first to gauge adequately the dimensions of the Aquarian frontier.

Rudolf Steiner's Akashic Rhapsody

Madame Blavatsky may be credited with having set the style for modern occult literature. Her contention was that a perennial wisdom lay buried among the world's cultural remnants and antiquities, where it waited to be found like so many scattered clues in a vast detective story . . . here, an amazing fact . . . there, an astonishing insight . . . and somewhere for sure, in the highlands of Tibet, in the Gobi desert, on a still-undiscovered Pacific island, in some forgotten corner of the world, a treasure trove of arcane lore protected by a secret fraternity of highly evolved *illuminati*. All that has changed about the style since her day is that the guardian wizards are now more apt to be located in outer space than in Shambala. Otherwise, the mythical pattern and philosophical assumptions remain the same: the path to the secret doctrine is omnivorous and freewheeling research in sources and authorities rejected by standard scholarship. All one need do is to look in the right places, get one's chronologies in order, apply the right interpretive keys to the folklore and legend of the world (usually by reading it as literal truth) and everything will at last fall neatly into place.

When we turn to Rudolf Steiner (1861–1925), one-time head of the German branch of Theosophy and later—after 1912—among the Society's leading schismatics, we encounter a radically different approach to occult studies, one that could hardly have been predicted from his early intellectual career. Until nearly his fortieth year, Steiner led the life of an accomplished if conventional academic, working toward his doctorate in formal philosophy, serving as scholar-archivist in the Goethe-Schiller Archives, editing a leading German literary journal, mixing intensely in the artistic and intellectual worlds of Vienna, Weimar, and Berlin. But then, after laying a studied epistemological foundation for his major vocation in life, Steiner struck out in a startlingly new direction as a "scientist of the invisible," drawing a wealth of esoteric teaching from direct visionary perceptions which had secretly preoccupied him since childhood. In effect, Steiner possessed those very scholarly and philosophical talents to which HPB had always aspired. And again, as with HPB, the emphasis of Steiner's worldview is on the evolutionary image of human nature; but Steiner's "occult science" (or "Anthroposophy," as he later called the movement he founded after being banished from Theosophy by HPB's successor, Annie Besant) abandoned literary materials and privileged communications with Tibetan *arhats*, in favor of projecting the pure "Akashic Record": the supersensible history of the universe accessible only to clairvoyant consciousness. What Steiner found in this "mighty spiritual panorama" was an evolutionary extravaganza no less complex than HPB's, and even more difficult to delineate, especially given Steiner's densely obscure style, an awkward mixture of heavy Teutonic system building and German Romantic *Schwärmerei*. It would be beyond our scope here to investigate his worldview in detail; but we can trace the major outlines of his evolutionary epic.

Because he wrote and lectured in a ponderous and syrupy rhetoric, Steiner was especially effective at suggesting huge cosmic vistas and the passage of immense periods of time, in particular the countless eons during which he imagines the architecture of the human psyche being fabricated by numerous echelons of supernatural agencies. His system begins in this cavernous past where nine hierarchies of spiritual beings—Seraphim, Thrones, Archangels, and so on—slowly weave a fourfold identity into the still-unconscious human personality, imparting to their unfinished creation their own spiritual qualities. What they toil at is actually a *co-evolution* among themselves, mankind, and the Earth, all of which are experiencing an intricate transformation and exchange of faculties. In Steiner's vision (as in HPB's) mankind and the Earth are portrayed as companion organisms which have been formed through a series of rhythmic "embodiments" and "spiritualizations" of the universe. During each of these oscillations, both human nature and the Earth have jointly matured and adapted to each other under the direction of the angelic intelligences, who are themselves mounting the ladder of evolution as they superintend the making of the human psyche. Thus, evolution governs all, cosmology as well as anthropology, and the basic theme of evolution is consciousness, from the angelic to the inorganic levels.

Thus far, in Steiner's system, there have been four oscillations or "epochs" through which human beings and the Earth have passed: the epochs of Saturn, Sun, Moon, and, currently, the epoch of the Earth itself. (Yet to come are the epochs of Jupiter, Venus, and Vulcan, all of which will be stages in a respiritualization of nature that will return the universe from its phase of material existence. Again, the parallel with HPB's secret doctrine is marked.) In each of these epochs, one of the four basic attributes of the human psyche has been slowly raised into existence along with a kindred form of consciousness in the natural world. Thus, we arrive at a series of suggestive, mystic correspondences between mankind and the cosmos, much in the tradition of the Hidden Wisdom's teaching that human nature is the microcosm of the universe. This sense of our weddedness to the Earth and to natural forces (which are always conceived of as beings) emerges as a kind of visionary neopaganism in Steiner, one of the more striking aspects of his Romantic heritage from Goethe. The correspondences are:

Saturn epoch (fire)	Sun epoch (air)	Moon epoch (water)	Earth epoch (earth)
deep uncon-sciousness, as in cataleptic trance	sleep conscious-ness, as in instinctive behavior	dream con-sciousness	wakefulness, self-conscious-ness
physical body	etheric body	astral body	Ego, personal identity
mineral realm	plant realm	animal realm	spirit realm

Interestingly, Steiner's system treats the physical body as the oldest and, therefore, most perfected aspect of our nature. Its perfection is such that it grows ill only due to interference by lingering instabilities in the higher levels of the personality: in the etheric body (the pattern of metabolic forces that governs the organism and which is susceptible to emotional upset), or in the astral body (the locus of self-awareness and willfulness), or in the Ego (the "individuated" self, as Jung would call it; the identity created by mankind's uniquely personal form of memory).

What Steiner sketches for us here is a vast, sympathetic network which relates the development of human nature to all levels of reality: to mineral, vegetable, animal, and finally, by way of the Ego, to the supersensible. Thus, human nature, from the bones up, recapitulates the evolutionary process and at last carries it forward into the realm of spirit. The primary function of reincarnation in Steiner's system is to elevate the lower levels of consciousness (which are, of course, *un*conscious in our present state of being) into full, mental vigilance. He envisages this evolutionary process as a metamorphosis within the organism itself achieved in the interim between death and rebirth. In this quiescent interval, for example, the angelic hierarchies slowly remold the person's metabolic organs (the physical seat of will and desire) into the head and brain of later incarnations, thus raising the "sleeping" instincts which we share with plant and animal to the level of conscious contemplation at the human level. All together, it is a compelling psychological image of our existential kinship with the whole of creation and of our embodiment of all varieties of consciousness. Steiner's conviction that the fully matured psyche must be conscious *down* through all its levels as well as *up* is one of the intriguing ideals of personal growth: to know the physical body to its marrow bone as well as we (at least potentially) know the heights of the spirit.

For Steiner, the appearance of the fully individuated personality does not mark the end of human development; beyond, in the next three episodes of evolution, lie the "stages of higher knowledge" that must inherit from the currently dominant empirical intellect. While these supersensible stages of consciousness—imagination, inspiration, intuition, as Steiner calls them—are not destined to mature until later periods of cosmic history, something of their quality is within reach of current humanity. These are, in fact, the levels of awareness Steiner himself claimed to employ in taking his occult science from the Akashic Record. His descriptions of the higher knowledge are subtle in the extreme and involve minute discriminations of experience that baffle ordinary thought. There is obviously no way to capture their complete meaning without pursuing the elaborate contemplative discipline Steiner created for teaching his students. His is one of the few systems of meditation native to the Western world, and for that reason alone deserves careful study and comparison with other traditions. Unfortunately, his followers have allowed this distinctive contribution to go undeveloped in favor of promulgating Steiner's writings and lectures.

The major work of the discipline is to strengthen faculties of supersensible consciousness that are now, in most of us, the helpless playthings of daydreams or hallucination. Steiner's exercises are meant to achieve a number of definite goals that will segregate and nourish these erratic powers. The objectives he sets forth include the stabilization of will and feeling so they can be deliberately directed, the increase of mental concentration and self-observation, and strict discrimination between arbitrary fancy and true vision. To achieve these ends, the bodily centers from which the supersensible faculties draw their energy must be systematically bolstered; the centers are described as "lotus flowers" and are located between the eyebrows, near the larynx, in the heart, and in the stomach—a remarkable, if incomplete, echo of Tantric cakra psychology. As the discipline approaches perfection, the student gains a new quality of sleeping and dreaming that contributes to a clear, objective experience of "spiritual facts" and finally achieves the liberation of a "second self" which closely resembles that "out-of-the-body" astral personality which shamans project into their visionary adventures. Steiner also insists, with commendable wisdom, that the ethical virtues of compassion and self-sacrifice must be exercised in tandem with the rising powers of consciousness as a corrective for the self-centeredness that may accompany mystic training. An independent Western discipline, Steiner's system is a significant confirmation of Eastern yoga techniques.

The overall evolutionary purpose of the three higher stages of knowledge is to strengthen the organs of "sense-free" experience which are mankind's means of entering the visionary realms. *Imagination* is the power to form meaningful and valid images of that realm independently of sensory perception, on the model of Goethe imagining the archetypal shape of the *Urpflanze*, the universal plant. A predominantly visual faculty, imagination penetrates to a "picture world" of forces and forms that are normally invisible. *Inspiration* is characterized as a kind of mystic "hearing," which introduces consciousness to the activity of spiritual beings in the universe; one begins to understand their history and interrelationships, which is, in effect, the occult science Steiner himself has already rehearsed in his writings. *Intuition* allows one to unite with these beings who govern our evolution, to know them and their work from the inside, as if their role were one's own. At this point, one begins to cross over into the future phases of the evolutionary process wherein

the whole universe . . . confronts the human being as a mighty edifice of Thought, even as the plant or animal world confronts him in the realm of the physical senses. . . . No sooner has he entered the supersensible world than he begins to perceive things that are the expression, not of anything physical, but of soul and spirit. These beings present themselves to him as an external spiritual world, just as stones and plants and animals present themselves to the senses in the physical world.[6]

But all this anticipates in individual experience phases of collective evolution yet to come. Meanwhile, our current Earth epoch must pass through a series of evolutionary subphases which take their origin from the calamitous fall of an ancient civilization that inhabited Atlantis. (Steiner follows HPB's story line of the Hyperborean-Lemurian-Atlantean civilizations, but his account is far less literal and reads like something more in the vein of psychohistorical fantasy.) In Steiner's myth, Atlantis is destroyed by the premature use of spiritual powers on the part of a still-impulsive, passion-ridden race. But the "initiates" of Atlantis survive the debacle of the black magicians and pass their knowledge on to a succession of later adepts: Zoroaster, Hermes Trismegisthus, Pythagoras, and so on. The "post-Atlantean" eras that ensue roughly correspond to major phases of the world's cul-

6. Steiner, *Occult Science* (London: Rudolf Steiner Press, 1969), pp. 280–81. Also see Steiner's *The Evolution of Consciousness* (London: Rudolf Steiner Press, 1966).

tural history. (The correspondence is *very* rough; the Akashic Record becomes sadly potted at this point and, disappointingly, includes none of the civilizations which the archaeologists have uncovered since Steiner's day.) The succession of eras is: ancient India, Egypt, Persia, Greece, the Christian West. Through all these eras, a single theme predominates: mankind's deepening involvement in physical matter and the progressive estrangement of consciousness from the spiritual. This is Steiner's version of the Hidden Wisdom's fall from grace. But Steiner gives an optimistic twist to this downward spiral of our evolution; like HPB, he welcomes it as a necessary stage in our destiny and in the redemption of nature as a whole.

Despite its alienative trend, the creation of the civilized arts and sciences is seen by Steiner as a worthy project; it has drawn out and elevated our powers of intellect, reason, and technical skill; it has led us into the intensive exploration of the world's physical properties. Above all, it has drawn the human personality out of its confused, instinctual participation in nature and given the individualized, self-aware identity the opportunity to develop. Yet, in the course of marrying itself to sensory experience and material reality, the personality has had to sacrifice its supersensible faculties until, in the fifth and sixth post-Atlantean eras (modern European culture), it reaches a state of toxic alienation. It is at this point that Steiner's occult science must intervene to teach mankind of the "spiritual powers that underlay the forces of nature." A major role in this turning of the evolutionary current must be played by a new, spiritualized Western science. Steiner remained to the last a faithful student of Goethe's *Naturphilosophie* in his hope that aesthetic and visionary experience would, in time, join empirical observation as integral parts of an etherealized objectivity—a worldview which regarded beauty, form, and meaning as *really* there, in nature, the message beneath the physical surface.[7]

While the evolution of human consciousness is, for Steiner, an activity in the charge of angelic hierarchies, there are also quasi-malevolent forces at work in the story. And, as always in Steiner's philosophy, these are conceived of as mythic beings, in this case the two tempters Lucifer and Ahriman. Although both are members of the divine hierarchies, they are fallen angels whose work is out of phase with proper human development. Their purpose is to pervert the destined course of evolution; but they also oppose each other in a complex way which allows the angelic hierarchies to play the two "side-streams" off against one another and so to appropriate their energy for good use.[8]

Lucifer and Ahriman make up a rich psychomythology in Steiner's work, one that deserves the attention of anyone who studies the evolution of consciousness. They are not an obvious duality. At the most general level, Steiner perceives Lucifer as the force that despiritualizes our experience of time and Ahriman as the force that despiritualizes our experi-

7. The best examples of Steinerite science are Theodor Schwenk, *Sensitive Chaos* (London: Rudolf Steiner Press, 1965), Hermann Poppelbaum, *A New Zoology* (Dornach, Switzerland: Philosophic-Anthroposophic Press, 1961), and Ernst Lehrs, *Man or Matter* (New York: Harper, 1958). Anthroposophists have also taken up Steiner's suggestion that the study of projective geometry should play a central role in reorienting our mathematical understanding of nature. See Olive Whicher, *Projective Geometry* (London: Rudolf Steiner Press, 1974). For a highly provocative example of Goethean science (non-Steinerite but in a style and sensibility Steiner would, I'm sure, have appreciated) see Lancelot Law Whyte, *The Universe of Experience* (New York: Harper Torchbooks, 1974).

8. Steiner's Lucifer-Ahriman mythology is contained in his *Three Streams in the Evolution of Mankind* (London: Rudolf Steiner Press, 1965). For a fascinating interpretation and application of the myth to several contemporary problems (sexual liberation, criminal justice, biological and physical theory) see Owen Barfield, *Unancestral Voice* (Middletown, Connecticut: Wesleyan University Press, 1965). Barfield is Steiner's best contemporary interpreter and probably the best place to begin a study of the Anthroposophical system.

ence of space. Time is parted by Lucifer from the timeless, and so becomes history; space is parted by Ahriman from the spaceless, and so becomes matter. Between the two of them, Lucifer and Ahriman would capture our consciousness in the wholly secularized realm of history and matter where all things spiritual would have to be reduced to purely physical and empirical causes. This project has made them central influences upon human development during the post-Atlantean epoch.

At a second level of manifestation, Lucifer is the rigorously conservative spirit of continuity and tradition who influences us to shun change and cling to the fixity of established forms. He seduces us by playing upon our sentimental attachments, seeking to subordinate innovation to strict heredity, an exact replication of the past. Over against him, Ahriman becomes the subversive spirit of revolution who seeks to obliterate the past. Instead of organic transformation, he would have the sudden, discontinuous substitution of new forms for old. In sociological terms, Lucifer is the spirit of all those who make heredity the dominant key to human personality; Ahriman is the spirit of all those who would totally remake the personality by manipulating its environment. In evolutionary terms, Lucifer would freeze the creative, mutational energies of man and nature and so produce stagnation; Ahriman would catapault consciousness prematurely into new phases of development for which it was not ready and so produce a caricature of evolutionary progress.

Finally, at the most personal and emotional level, Lucifer is the spirit of sentimental warmth and pity, who works to defeat even the most salutary change in our lives—rather in the way Dostoevsky's Grand Inquisitor perversely sought to negate the second coming of Christ out of loyalty to the Church and a twisted compassion for mankind. All those who would seek to re-create ancient and primitive religious forms on the Aquarian frontier would, in Steiner's view, be falling prey to Lucifer's sentimentality. In contrast, Ahriman is a cold and angry spirit who appears in our feelings as impatience, power-lust, and domineering will. He is peculiarly the spirit of scientific intellect and aggressive "progress," as well as being the rebellious indignation that underlies the secular ideologies of our time. All those on the Aquarian frontier who would ground spiritual development in laboratory findings and physiological research, as if the past had nothing to teach us, would be the children of Ahriman.

Thus, both spirits work upon necessary and healthy emotions—but do their work one-sidedly. We need their energy, but it must be synthesized if we are to achieve a transformation of consciousness, rather than either stagnation or abrupt substitution. Where there is no synthesis and both spirits have their way with us, the effect on our most intimate experiential powers is to destroy our capacity to see the spiritual meaning of birth (which is always rebirth for Steiner) and of death, to isolate each person's biography from its preexistent and posthumous development. Ahriman works against our memory of spiritual preexistence (our spiritual heredity); Lucifer works against our perception of the eternal beyond death. But the two tempters never quite achieve their ends, for evolution is a *three-way* process, and the third force in evolution—which is the divine plan pursued by the hierarchies and finally championed by the Christian logos—balances their intentions and synthesizes their power into the evolutionary mainstream. Thus, Christ's miraculous birth and his miraculous resurrection reveal the spiritualized conception of birth and death which Lucifer and Ahriman seek to obscure.

The distinctive aspect of Steiner's evolutionary system is his insistence that the spiritualizing turn which evolution is destined to take will not be a mere *re*turn to the past; it will become a new era in which we integrate all we have known and been through the many incarnations of the personality. Nothing has happened in vain, no era of history, no form of consciousness—not even those generated by Lucifer and Ahriman, for even these ultimately serve to introduce us to experiences of time and matter that will finally be ethereal-ized. In reality, Lucifer and Ahriman have unwittingly exer-cised our capacity for free decision, and out of our struggle with their influence a new, sacramental unity with nature will finally be evolved. In that unity, the human Ego will be perfected in its personal identity; in contrast with what Steiner understood of the Eastern religions and their version of reincarnation, the personality will not be blended out of existence or lost in the godhead. Rather, it will take its place in the ranks of the spiritual intelligences as a new form of being. For that is what our entire evolutionary adventure is all about: the creation of the tenth angelic hierarchy, the Spirits of Freedom.

There is a strong Romantic impulse to Steiner's high re-gard for personality which echoes Goethe and Nietzsche, though in a more mystic tone. Unfortunately (or so I find it), Steiner saw fit to affirm this central idea by making Christian-ity and the "mystery of Golgotha" the axis of his entire sys-tem. It is ultimately the "Christ-event" that sways the great course of evolution, and the Christian conception of soul that dominates our evolutionary fate. So in the Goetheneum, the building which Steiner designed to serve as the world head-quarters of Anthroposophy in Dornach, Switzerland, he planned a sculptural group to serve as the focus of the struc-ture: three figures carved in wood—Lucifer, Ahriman, and Christ.

Although Steiner's Christianity is unorthodox to the point of heresy and his interpretation of scripture among the most eccentric in Christian history, the commitment gives his philosophy a glaringly parochial character—and it was just this ethnocentric reversion which led to his falling out with the Theosophists.[9] It also forces him into some sadly distorted readings of the Eastern religions, which he negatively stereotypes as irresponsibly as any Victorian missionary. Still, in Steiner's celebration of the etherealized personality we have one of the strongest defenses of the Christian soul—that deathless, personal identity which so many modern therapies and mysticisms have vilified as a diseased illusion. What Steiner invites us to contemplate is a liberated individuality on the far side of neurotic individualism, a personal consciousness which, though born of alienation, can yet become the vehicle for transcending our isolation from nature and can carry us into a new communion with the whole.

Regrettably, Steiner, who lavished a surplus of lush Germanic prose on his every visionary experience, was hardly the one to appreciate the reticence with which Eastern masters like the Buddha, Lao Tzu, and the Upanishadic philosophers surrounded their highest insights, nor to see the promise of a nonegotistic personality that might be hidden in their wise silence.

Gurdjieff's Therapy by Ordeal

For Madame Blavatsky, the evolution of consciousness was a secret doctrine to be gleaned from the dusts of tradition and from privileged sources of instruction. For Steiner, it was a cosmological process to be recaptured in direct, visionary experience. For George Gurdjieff (1872–1949), the third major occult evolutionist of the past century, that image was, as were all ideas for him, a useful tool in the hands of a psychic technician. While Gurdjieff also swaddled his thought about evolution in a grand cosmic design that drew in the planets and the elements, his strength was

9. Though at the time of the split, the Theosophists under Annie Besant's leadership were involved in their own rather embarrassing parochial detour to the East. Mrs. Besant was then touting the young Krishnamurti as the messianic Buddha.

not that of a systematic metaphysician. Perhaps he had learned from the Sufi masters, whose disciple he claimed to be, the virtues of silence and indirection when dealing with the great teachings. In any case, his was that sort of practical, pedagogical mind that has more use for a parable or an instructive anecdote than for a poetic flight of the imagination. Like HPB and Steiner, Gurdjieff also draws upon the Atlantis motif; but it becomes for him much more a didactic fiction than a historical assertion or a visionary myth. As for his cosmological ideas (the law of octaves, the law of three, "feeding the moon," the seven cosmoses), they are meager and poorly developed snippets, some of which remained bewilderingly obscure even to his closest students.

There have been some heroic efforts by Gurjieff's disciples, especially P. D. Ouspensky and J. G. Bennett, to salvage coherent philosophy out of Gurdjieff's teachings. But one simply cannot tell, in reading them, where Gurdjieff leaves off and where they take over; their systems are not all together compatible with one another. When Gurdjieff himself put pen to paper late in his career to write *Meetings With Remarkable Men* and (his major work) *Beelzebub's Tales to His Grandson,* what he produced was done in a perversely oblique, often comically overblown style whose general obscurity suggests he wished to keep more secret than he revealed, always a weak foundation for literature. His books teem with tall tales, interminable circumlocutions, and madcap polyglot neologisms ("Foolasnitamnian", "common-cosmic-Ansanbaluiazar", "Heptaparaparskinokh"). In truth, Gurdjieff's spokesman in the Beelzebub tales is a great, cosmic windbag, amusing at first, but soon enough a tedious bore. No doubt the writings are crammed with "logominisms" (Gurdjieff's made-up word for secret doctrines), and perhaps, as one follower has advised me, there is no grasping their deep meaning until one has reread the master at least forty times. Unfortunately, the writing lacks the intellectual magnetism that invites such intense study; and when the doctrines are spelled out by loyal disciples (we will investigate a few below), they hardly warrant the surface obscurity that is supposedly guarding them. Certainly the heavy tongue-in-cheek tone throughout makes one wonder how much of all this is a put-on. There is rather too much of Baron Munschhausen in Gurdjieff the writer.[10]

The overall impression one forms of Gurjieff's work is that cosmology and metaphysics, in so far as he used them at all in his teaching, figured as mere background sketches for his main concern, which was to revolutionize the personality by direct, therapeutic intervention. Gurdjieff was a man of technique, of discipline, of *praxis*, working out of a strong vision of psychic pathology and health. His evolutionary ideas apply most significantly as a model for guiding the individual soul in its struggle for liberation. His major contribution, as one of the West's pioneer gurus, was to harnass the evolutionary image to a repertory of educational innovations in which we can now recognize the seeds of many contemporary eupsychian therapies: T-groups, encounter groups, Transactional Analysis, Synanon games, Erhard Seminar Training, and so on. All are profoundly indebted (largely without realizing it) to the strange experiments in human relations and personal growth which Gurdjieff inaugurated at his Institute for the Harmonious Development of Man at Fontainebleau in France during the early 1920s. His goal was never really to record metaphysical doctrines, but to lead students to the experience of personal discovery. The rest was—or might as well have been—silence.

Like HPB, Gurdjieff loved to surround himself with an atmosphere of impenetrable mystery. A Russian-Cappadocian Greek emigré of obscure origins, the life story he

10. But, in fairness, one should consider J. G. Bennett's defense of Gurdjieff's style and his exposition of the major doctrines in *Gurdjieff: Making a New World* (New York: Harper & Row, 1973), by far the best introduction to Gurdjieff as teacher, philosopher, and personality.

carried to Western Europe with him following the Russian Revolution implied that he had studied at the feet of secret masters in the Orient. His habit was to attribute his teachings to ancient fraternities of the middle east with whom he supposedly remained in close touch, perhaps as a special emissary on a mission of Earth-shaking importance. He spoke much of an "inner circle of humanity", whose confidant he was. The claims are somewhat more credible than HPB's, but nearly as difficult to substantiate. So much seems clear: the cultural melting pot in which Gurdjieff grew up (Armenia, Persia, Turkestan) has harbored secret societies of "Kwajagan" (masters) since long before Islam came into the region. Gurdjieff, for example, made oblique references to membership in a Sarman Brotherhood which ran back to the days of Zoroaster. J. G. Bennett, who has done the most conscientious job of examining these claims against their historical background, finds them plausible, believing there may even be an unbroken acroanomic tradition in the area reaching back 30,000 years to ancient shamanic fraternities whose lore has absorbed many later religious cross-currents (Buddhism, Christianity, Islam) and has been passed down to such extant groups as the Naq'shbandi dervishes. Bennett finishes his examination convinced that Gurdjieff was the student and agent of some such spiritual elite.[11]

Unhappily, again like HPB, Gurdjieff combined these exalted credentials with an impish taste for mystification that cannot help but arouse suspicion. His general manner has

11. Again, the best place to review Gurdjieff's claims and possible connections is Bennett's *Gurdjieff: Making a New World*. Curiously enough, Rudolf Steiner's interest was also drawn to the turbulent middle-eastern cultural complex where Gurdjieff began his search for wisdom, but from a very different perspective. Steiner locates in this area, at the Persian Academy of Jundí Sábúr founded in the sixth century A.D., an anti-Christian, wholly "Ahrimanic" brains trust which would apparently have invented modern science a thousand years too soon and would have disastrously derailed human evolution. (See Steiner's *Three Streams in the Evolution of Mankind*, London: Rudolf Steiner Press, 1965, chapters 5 and 6.)

been described by Ouspensky as that of a man "badly disguised" and *self-consciously so*—as if to keep those he met off balance. He was never forthright about his background, even when dealing with sincere, intelligent people who clearly deserved his trust. Moreover, he often played the charlatan and obfuscator deliberately, and delighted in the part. He boasted of the cunning business deals by which he again and again solved "the material question" (as an oriental antique and carpet merchant, for example) and frequently called upon his students to help him stage phoney magic shows for the public. He was a gifted hypnotist capable of using his powers to cure drug addiction, but also prone to assume a Svengali-like pose with his more impressionable pupils, especially the women he attracted. There is an air of the old shamanic trickster about him, as well as of the loveable Falstaffian rogue, replete with great bankrupting appetites for fine food and drink. While I confess to having a personal weakness for such colorful characters and admire their brash resourcefulness, I am not among those who find such qualities encouraging in people who purport to be spiritual masters; in these, I prefer the utmost candor and transparency. Perhaps I have been exposed a few times too often (and to my regret) to mystagogues equipped with fearsome, penetrating eyes and the seemingly clairvoyant power to probe one's very soul. There are those who can amaze without limit, but never once enlighten.

Yet, in Gurdjieff's case, there are special and redeeming circumstances. These lie primarily in his talents as guru and therapist, particularly in the techniques he developed for affective education. Bennett believes the key elements in Gurdjieff's methods were taken directly from Sufi masters. These include the use of the communal group as a setting for spiritual work, various exercises for intensifying concentration on the here-and-now, the tactic of dogged, ruthless encounter with the student's every lie, weakness, and evasion, the calculated use of shock and surprise. If this is so, then much of modern eupsychian therapy may have been borrowed, by way of Gurdjieff, from ancient schools of middle eastern spirituality.

At the heart of Gurdjieff's therapeutic work lies the assumption that "man is specially made for evolution—he is a special experiment made for self-development." [12] But Gurdjieff strictly distinguishes human evolution from biological (or "mechanistic") evolution, which he generally treats with aristocratic contempt. Evolution, he insists, is not mere randomized change; it is "conscious, voluntary, and intentional development of an individual man on definite lines and in a definite direction during the period of his Earthly life." Evolution is the result of a deliberate action of the will; it happens no other way. Anything else is drift, flux, a meaningless to-ing and fro-ing. Evolution belongs only to the human animal—and not even to *all* of these. Most people never achieve consciousness at all, but remain in a sleeping condition. They are no better than machines hopelessly in the grip of unexamined, uncontrolled, external influences. And "nothing," Gurdjieff argues, "evolves mechanically. Only degeneration and destruction proceed mechanically." Evolution is for the few, the "inner circles," the "conscious nucleus" which has awakened to its full human identity. These few alone deserve to be called "humanity"; the rest are so many automatons, and on them Gurdjieff wastes little compassion and less time. There is little of the bodhisattva's mercy in him.

For Gurdjieff, the evolution of consciousness is an exceptional event, a "super effort" which nature does nothing to facilitate. Indeed, "the way of development of hidden pos-

12. I am guided in my comments on Gurdjieff by the reports of P. D. Ouspensky, especially as carried in his *In Search of the Miraculous* and *The Fourth Way* (New York: Harvest Books, 1949). I use Ouspensky's and Gurdjieff's words interchangeably here.

sibilities is a way against nature, against God." Nature's interest is in the mechanical, which it binds and stifles with laws, rules, external determinants. Evolution fights against the grain; it is *hard work*. The emphasis on struggle and defiance in Gurdjieff's teachings is unrelenting and lends a drab, strenuous tone to his thought. Surely no therapist has ever had a grimmer vision of how difficult it can be to decondition people and free their spontaneous, self-critical nature.

What is most striking in Gurdjieff is his notion of what authentic consciousness—and so authentic evolution—is. Like Steiner, Gurdjieff worked from a fourfold psychology: carnal body, astral or natural body, mental or spiritual body, causal or divine body. The mechanistic many possess only the carnal body, and even over this they hold no reliable control, helplessly obeying its every stimulus and distraction. They move in the "vicious circle of automatism," victims of "mass hypnosis." But as one struggles through one's evolutionary course, one acquires the power to police each body in turn and so to grow into the next. Each step in this evolution is marked by progress in self-observation and self-control. To evolve the "causal body" (the body which "bears the causes of its actions within itself and is independent of external causes") is to free oneself from all erratic, emotional determinism: bad temper, annoyance, fear, anxiety, social role playing, competition, pretense. At the level of the causal body, we achieve unshakable and autonomous individuality, free from every form of conditioning. As Gurdjieff put it in one of his semicomic circumlocutions, we gain the "potency-not-to-be-identified-with-and-not-to-be-affected-by-externals-through-one's-inevitably-inherent-passions." In this way, we at last create an "I" which is permanent, obedient, and of our own choosing. We unify all the "centers" of the organism (intellect, emotion, will) so that none can be taken over by outside forces and used to unsettle us.

In his Beelzebub tales, Gurdjieff worked out a cosmic setting for the human evolutionary struggle. In the spirit of fable or parable, he suggests that some 100,000 years ago the higher powers of the universe determined that human evolution had to be slowed down because it was somehow endangering the equilibrium of the solar system. So they intruded an organ called "Kundabuffer" at the base of the protohuman spine which cooled off the evolutionary pace for a time. The word and the idea are clearly borrowed from the kundalini of Indian yoga. But oddly, or perhaps ignorantly, Gurdjieff understood the kundalini to be a treacherous obstacle to enlightenment. Then, about 40,000 years ago, the organ was removed and modern homo sapiens was allowed to evolve. But, unfortunately, many of the Kundabuffer characteristics continued to exert their influence over our species, making human beings egotistical, conceited, and subject to constant distraction by illusions and trivialities. As a result, the main role assigned to mankind in the universe—that of "Reciprocal Maintenance"—has been disrupted. We are no longer reliably serving as "an apparatus for the transformation of energy" from lower to higher levels. As J. G. Bennett interprets this teaching, mankind, as energy transformer, plays a crucial part in the evolution of the cosmos as a whole, in conserving the balance of nature, in protecting the well-being of plants and animals and the health of the soil. The entire seven-level hierarchy of cosmic being is in jeopardy of disorientation due to the human loss of right consciousness.

Bennett, who was among Gurdjieff's last students and who had the benefit of private conversations with the master, has managed to assemble a coherent cosmology around the doctrine of Reciprocal Maintenance. It is more than I think most readers would find in Gurdjieff's odd fable of the Kundabuffer. Bennett sees in Reciprocal Maintenance a precocious vision of our ecological responsibility to the Earth. The teaching may be that, but it is surely a cumbersome and overwrought way of coming at the ideal; it is simply more metaphysics than we need to recognize the environmental facts of life, and not very engaging metaphysics at that. As with Ouspensky's efforts to elaborate Gurdjieff's fragmentary teachings, the result is an angular, overly-complicated system in which what is most novel is most uninteresting.

In any case, the Kundabuffer fable makes so much clear: Gurdjieff shared the conviction of the Hidden Wisdom tradition that the evolution of human consciousness is intimately related to the evolution of the cosmos at large; self-transcendence is therefore the only solid basis for moral discrimination. Striking a Nietzschean note, Gurdjieff places evolutionary destiny beyond good and evil.

> The only possible permanent idea of good and evil for man is connected with the idea of evolution. . . . If a man understands he is asleep and if he wishes to awake, then everything that helps him to awake will be *good* and everything that hinders him, everything that prolongs his sleep will be *evil*. . . . Good and evil exist only for a few, for those who have an aim and who pursue that aim.[13]

Like Steiner, Gurdjieff defends the rock-bottom reality and paramount value of the person. He anticipates no immersion in the godhead; the personal soul stands firm and endures. Indeed, only the fully conscious individual can expect immortality, with perhaps some form of life "recurrence." The others perish utterly. Gurdjieff deviates from the Hidden Wisdom in treating reincarnation as a mere "theory" worth little investigation.

13. Ouspensky, *In Search of the Miraculous* (New York: Harvest Books, 1949), p. 158. What this ethical pronouncement omits, of course, is the remarkable moral consensus on the part of the world's saints and sages, who seem quite spontaneously moved to endorse the values of gentleness, compassion, love, and service. Who are we lesser mortals, one wonders, to pretend that perhaps *our* enlightenment (if it ever comes) may call all this into question?

Gurdjieff refers to the fully awakened state as "objective consciousness." Unfortunately, he does not succeed in making this a very appealing state of being, at least at the distance from which he describes it to his pupils. To be awake appears to be a status that knows no tranquillity, easiness, or grace. Rather, it is a constant labor dedicated to observing and controlling the self, beating down all spontaneity and joy. Although he was personally gifted with a strong comic sense (it shows up in the superloquacious zaniness of his Beelzebub tales), Gurdjieff allowed little humor to flow into his therapeutic work, perhaps for fear of breaking down the discipline of the group, where intensely critical, utterly sober self-awareness was the rule. In the groups, people *toiled* at their evolution, often by the sweat of their brow. Hard and heavy labor, exhausting work assignments, severe physical and emotional testings were prominent techniques. And, too, the sort of therapeutic abuse and irritation that were later to become the commonplace (and often the cruel commonplace) of encounter groups and Synanon games. "What is necessary to awake a sleeping man?" Gurdjieff asks. "A good shock is necessary"—and one never knew when such a timely shock might be applied by the master.

One sees in this method something of the stern discipline and surprise attack methods of Zen and Tibetan masters, again with the same purpose of cultivating "mindfulness" to the here and now. To this end, Gurdjieff invented his legendary "movements," intricate group dances that require a rigorous coordination of limbs and exacting teamwork. Supposedly, the dances were adapted from the "sacred gymnastics" of the Sufi communities. Ironically, the result is a set of extremely rigid choreographies so mechanistic that one almost concludes Gurdjieff intended them as a clever ruse for his students: who among the dancers would prove "awake" enough to disrupt the tyrannical pattern and go their own free way? But no; the dances were meant to inculcate eternal vigilance, constant awareness of oneself in the action of living. So too the "stop exercises," which required students to freeze in position without warning and then to reflect on *this* moment, *this* action, *this* feeling—right now, right here. At times, the image of the fully enlightened personality we have from Gurdjieff almost seems to be that of a man who always knows what his left foot is doing.

For all the criticisms one might make of Gurdjieff's often heavy-handed and humorless methods, we have in his work the first Western effort to ground psychotherapy in the evolutionary image. With Gurdjieff, the purpose of therapy was not to excavate the causes of neurosis, but to open the latent, higher centers of the mind, to evolve the personality forward. He was among the first in our time to bring therapeutic work to the essentially well person as means of further growth. He took therapy out of the grubby basement of old guilts and fears and lifted it to the penthouse of new potentialities. This is, in fact, a figure of speech which Gurdjieff frequently employed: the psychological house we inhabit has an upstairs; but we do not know the upper story is there. That is the task of eupsychian therapy: to lead us up the staircase.

Gurdjieff also offers an important caution to those like Teilhard de Chardin who find in the evolutionary image a guarantee of limitless human progress. For Gurdjieff, evolution is not cosmic law; it can fail and go wrong, and most often does. Waking up is a supreme effort of the individual will, following careful instruction and guidance; it comes about in no predestined way. It requires a strict master, "deliberate suffering" (in Ouspensky's words), and a supportive group of fellow students. But where there is the possibility of evolution, there is also the possibility of devolution, decline, degeneration. And of this Gurdjieff saw enough all about him in this world of "mad machines" to touch his work with a despairing resignation. If he was one of the most innovative therapists of our time, he was also one of the most cheerless.[14]

The Evolution We Deserve

In the work of Madame Blavatsky, Steiner, and Gurdjieff, we can see a steady progression of the evolutionary image from exotic theory to therapeutic practice. With HPB, the image remains embedded in an abstruse and intricate metaphysical system that borrows heavily from ancient and alien sources. It was not a style that was apt to find a graceful place in the Western intellectual mainstream. We can credit HPB with having sought out the origins of evolutionary thought in the Hidden Wisdom and having shown with an admirably disputatious spirit how widely modern biology has deviated from that tradition; but her work continues to be a quaint monument that our culture is not likely to assimilate. What was most durable in it—her explorations of the still-living religions of India and Tibet—has by now become available and widely familiar from any number of knowledgeable modern sources. Indeed, the lamas and the swamis themselves are now among us to offer their lore and techniques at first hand.

With Steiner, the evolutionary image becomes decidedly more psychological in character. The allusions and vocabulary of his system are still odd to the point of being jarring; and there is still, as with HPB, a determined attempt to systematize far more than any one mind could ever convincingly claim to have mastered. But Steiner's emphasis is distinctly introspective. The research we are presented with is not, like HPB's, an exercise in scholarship; it has become

14. For a shrewd, slashingly negative critique of Gurdjieff, see Whitall N. Perry's series "Gurdjieff in the Light of Tradition" in *Studies in Comparative Religion* (Bedfont, England) Autumn 1974, Winter 1975, Spring 1975. Perry makes mincemeat of Gurdjieff's metaphysics and comes close to characterizing him as a diabolical agency.

internal research, a matter of searching out the evolutionary forces in the depths of one's own experience. We have moved away from reliance on HPB's "very old books" and secret gurus toward personal, visionary discovery. As a result, Steiner produces a psychomythology which is far more mature than anything we find in Freud or his disciples; for in Steiner's myths of the mind—especially in such visionary beings as Lucifer and Ahriman, Gabriel and Michael, the angelic hierarchies and the guardians of the psychic thresholds—we find the superconscious as well as the subconscious taken into account. Most important, Steiner's work includes a well-defined, contemplative discipline for the pursuit of "the higher knowledge." Perhaps one could also include here some passing mention of Steiner's movement-art Eurhythmy, which, like Gurdjieff's dances, represents a pioneering effort to develop a nonverbal mode of education. Unfortunately, for all his emphasis on meditation and personal discovery, Steiner could not restrain himself from promulgating a total and densely detailed world view which includes fully articulated systems of medicine, "biodynamic" agriculture, art, and education. Steiner was one of those leaders who leave their followers very little more to do than to vary the master's themes and to express endless gratitude. As a result, the Anthroposophical movement, since Steiner's death, has been all but weighed down to the point of immobility by his authority on all matters and has too much contented itself with secondhand exegetical studies in his voluminous writings. Still, the spirit and clear intention behind Steiner's work was to ground occult science in the personal experience of students.

Finally, with Gurdjieff, we cross the boundary between metaphysical discourse and psychiatric practice. If everything we have of Gurdjieff's cosmology and metaphysics were to be lost, his importance would be fully preserved in his therapeutic techniques and in the evolutionary model that governs them. In psychiatry Old Style, as it was developed by Freud, therapy borrowed its objectives from medical science; it was content to patch up psychic lesions and moderate the virulence of the instinctual forces. But following Gurdjieff, therapy moves into the lives of reasonably well and functioning people whose "illness" is not neurotic disability but unrealized potential. In this work, therapists become more in the way of gurus rather than psychic engineers, and the model they apply in their search for sanity is one that describes spiritual powers that have yet to be unfolded by a disciplined action of the will. This is the model that has been taken over by the Human Potentials movement from Gurdjieff's work, along with numerous versions of his group and encounter methods for breaking down the ego defenses to liberate what lies behind them.

So it is in the form of therapy that the Hidden Wisdom, like so many of the Oriental traditions to which it is subtly related, has significantly entered the modern world, and with it the sense of evolution as our destiny in the making. *Transcendent potentiality:* that is the core of the Hidden Wisdom's drama of cosmic transformation. In that one idea, we can include the three concepts which Darwinian biology has exiled from scientific thought: value, intention, destiny. From occult tradition, by way of Blavatsky, Steiner, and Gurdjieff, we inherit the pregnant notion that evolution has an experiential *inside,* an autonomous and guiding thrust with which we can ally ourselves. That idea is the beginning of the great transformation that awaits us. For as evolution passes over to the level of consciousness, *our understanding of evolution in itself becomes an evolutionary factor.* We undergo the development we envision for ourselves; we get the evolution we deserve. If we continue to see evolution as an empty game of chance in which will and aspiration play no part, then we doom our own higher development. We will, in effect, have *willed* ourselves into impotent drift and stagnation. If we recognize evolution as the unfoldment of visionary energies, then we will have liberated those energies as an evolutionary force, and not only within our own lives, but within the history of our species as a whole. Our situation is perhaps very like that which prevailed at the time language made its appearance among our human and protohuman ancestors. To begin with, in that first generation of talkers, the concept and practice of speech must have had to be unfolded by individuals out of their own mental potentialities. But the spiritual environment of the time must have been saturated with language-readiness, like a seeded field whose germination has come into season. One person's act of speech called forth another's latent talent, and the gift of language blossomed throughout the species to become a universal defining characteristic. So too the cultural moment we inhabit may be soaked through with visionary talents soon to mature on a planetary scale.

7

Triangles, Sacred and Profane:
The Visionary Origins of Culture

> The great Adept may indeed have to hide much of his deepest life, lest he tell it to the careless and the indifferent, but he will sorrow and not rejoice over this silence, for he will be always seeking ways of giving the purest substance of his soul to fill the emptiness of other souls. It will seem to him better that his soul be weakened, that he be kept wandering on the earth even, than that other souls should lack anything of strength and quiet. He will remember, while he is with them, the old magical image of the pelican feeding its young with its own blood; and when, his sacrifice over, he goes his way to supreme adeptship, he will go absolutely alone, for men attain to the supreme wisdom in a loneliness that is like the loneliness of death.
>
> —William Butler Yeats

The moment of visionary ascent I have mythically attributed to the Few is a private adventure, a fire in the depths of the charmed personality. It hides in the reluctant silence of Yeats' great Adepts, who must, in spite of themselves, surround their knowledge with the loneliness of death. Yet such moments are the brightest human experience of transcendence, and because that experience lights the path of our evolutionary development, it must in some fashion be shared with the needy world roundabout. Either that, or the visionary energies stand no chance of becoming a saving historical force. How close, then, can we come to that visionary fire, those of us who are among the many and not the Few?

The Habit of Duality

To ask this is to ask what the relationship is between the sacred and the profane. And that is not a question our society is suited to answer gracefully. We, peculiarly among all peoples, balance between extremes of being that cruelly divide the integrity of our lives. Heaven and hell . . . body and soul . . . nature and grace . . . mind and matter . . . our heritage is a repertory of crazy-making dichotomies. Like these, sacred and profane are also for us a radical disjuncture, a forced choice between absolute incompatibilities. This habit of duality, of tearing our personality between the light and the darkness, comes built in with our religious heritage. The god of the Jews and Christians is One; but he is not (like the pantheist's god) *everything*. Nature is neither his body nor his garment, but an object of his making. He does not permeate the world like its soul or substance; he rules and upholds it, standing over and away from his creation in lordly dominion. And there is much in that world he sets his face against. A lawgiver must, after all, have his outlaws.

Even when, in the Christian adaptation of that tradition, god becomes flesh, the miracle is cut down to human size and monopolized by one of our lesser needs: the need to escape the unwholesome shame we heap upon our creatureliness. The divine act of incarnation is not interpreted by Christians as a symbol of nature's participation in divinity; that must be rejected as pagan idolatry. Rather, the incarnation is seized upon as the promise of pardon from an insupportable burden of sin. Unconfessed angers, sexual guilt, the promptings of pride and rebellion—all these burden our natural animal energies, making them a source of self-loathing and turning death into a punishing misfortune. Beneath their weight the savior's passion ceases to be what it was in the mystery cults of the ancient world—a celebration of the Eternal Return in nature, as it has been known and celebrated since time out of mind. Instead, it is scaled down to a special act of redemption; it is historically bounded, literalized into a unique event, above all *moralized* into a blood sacrifice to solace the anxiety of guilt-ridden souls. It is a question no Christian theologian has been willing to address: how many believers have craved the blood of the lamb for nothing better than to wash away the secret guilt of masturbation fantasies?

Thus delimited by trivial and unbecoming needs, the Christian incarnation leaves us at last with a "body of Christ" which is not the world at large—sky and sea, star and mountain, bird and beast—but only *the church,* the congregation of accredited believers carefully segregated by their confessional correctness from pagan, heretic, and worldling. The Kingdom remains a walled city; beyond lurk sin and the infidel, the unclean and the demonic. Stand at the boundary and *choose.* The one side or the other.

Either/or. And between the *either* and the *or* there stretches Kierkegaard's "infinite qualitative difference," the absolute divorce of sacred from profane.

Perhaps in their workaday routine, few people realize how much of their best energy is steadily, secretly drained away by this schizoid tension, this psychic fault line that runs down the middle of their lives. But every dichotomy this culture clings to forces us to choose, and every choosing is a repressing, the exile of some outlawed part of ourselves. We live always between the devil and the good lord. No matter how we rename the contestants that wrestle for our soul (reason against passion, ego against id, Eros against Thanatos, anima against animus, bad faith against authenticity, sanity against madness), always, at last, it is ourselves against ourselves across the battleground of the divided soul. Unfinished an animal as we are, we diminish ourselves still further by this lifelong exorcism of demons that are, in reality, so many fragments of our own forbidden experience.

To choose *either* half of any metaphysical or psychological dichotomy is only to fall into the same trap of self-divisiveness all over again. *The dichotomy is the problem.* That is what forces us to split and repress the self. One sure generalization we can make for very nearly all the psychic and spiritual experimentation that fills the Aquarian frontier: its purpose is to close the dichotomies that have for so long tyrannized Western society, to reunite flesh and spirit, intellect and instinct. Above all, it seeks to bridge the rift that parts sacred from profane. Its search is for the Earth's epiphanies, for *experienced* divinity, in the world, here and now: the gold that lies hidden beneath the dross of doctrine and dogma.

"Recognize everything around you as Nirvana; hear all sounds as mantra; see all beings as Buddhas." So John Blofeld was instructed by his Tantrist lama.[1] That sacramental vision of the world has been most faithfully preserved in the surviving primitive religions of the world and in the mystical traditions of the East. But it is also the sensibility which underlies the several schools of the Hidden Wisdom, and which most sharply separates them from the Western religious mainstream. For the Hermetic philosophers as for the Tantric Buddhists, the body itself was a magical object linked mystically to the planets and elements of nature. For the Western alchemists as for the Taoist sages, all nature was a mirror of the divine. For the Kabbalists, every letter and number invented by the mind was a transcendent symbol that cast shadows of the sacred across our Earthly life.

In all these traditions, as much as they may differ in style

1. John Blofeld, *The Tantric Mysticism of Tibet* (New York: Dutton, 1970), p. 76.

and technique, the same central teaching is honored. The Hermetic philosophers expressed it in a famous formulation: "As above, so below." Find the sacred within the profane, the profane within the sacred; for the two are organically related, hierarchically ordered. Or so they are experienced and taught to be by the visionary Few, whose role is to keep open the bridges of experience that run between the two realms. That, as I have suggested earlier, may have been the original function of culture: to provide the many with the symbols, images, and rituals that would allow them to participate in the knowledge of the Few.

Theme and Variations

If the sacramental unity of sacred and profane was the primordial subject matter of culture (and so I will assume here) then there is a sense in which culture-making must always be a sacred activity, for it must always work with some residue of that vision. But that, obviously, is not how most people in our society experience the culture they live in. Rather, they know it as a totally profane province of workaday things, sordid affairs, frivolous activities wholly untouched by anything divine. They know it as the spiritually void world we have "fallen" into, and whose concerns are the very opposite of the sacred. We either resign ourselves bravely to its fallenness or suffer it as the curse of our human condition, until, at last, the very awareness of any reality besides the profane fades from our lives.

And that is the paradox of culture, especially as it has unfolded in Western society. Culture begins as the signature of the sacred traced upon the mind by visionary awareness; but it is culture which finally alienates us from the sacred by building the profane world about us.

As my myth would tell the story, then: culture is what the

Few create in order to confer their vision upon the many. Image and gesture, word and symbol, originate as their light-bearing media. But once created, culture takes on a life of its own. It reverberates from mind to mind, and as it does so, it *evolves*—out and away from its sacred beginning toward new possibilities. We playfully improvise with the elements of culture, bending and shaping them to suit all occasions, from inspiration to entertainment, from ecstasy to utility. At last, what we call "profane" is only that portion of culture in which, after long use, we can no longer clearly see the visionary connection. *The profane is a sacred theme varied beyond recognition.* But the connection is still there, because, even in its most trivial or grossly utilitarian mood, the mind has nothing else to make culture out of except the remnants of transcendent insight.

I return to the image of evolution as an unfolding destiny: growth from a seed. Within every seed, there hides a genetic paradigm, the finished form waiting to be displayed. It is all there, but hidden. Though wholly unexpressed, the whole is self-sufficiently within, contained in that paradigm like an idea in the mind that has not yet put on language. Whatever the waiting paradigm eventually unfolds is naturally and harmoniously itself, all of it, every part and piece. Roots do not contradict branches, though they grow in opposite directions; the fruit is not at odds with the flower, though the one is born from the death of the other.

That is the evolutionary unity of sacred and profane we must come to see in culture: in Lancelot Law Whyte's phrase, a "unity in process." We need to see the profane as the creature of the sacred, just as rough Caliban was the beloved, if sometimes unruly, creature of Prospero. And we need to see that fathering act happening purposefully in history as part of the human culture-making project. Because only in this way can we prevent history (which is our collective human biography on Earth) from becoming the antagonistic opposite of eternity, and so saddling ourselves with still another tormenting dichotomy.

Culture Upside Down

Again the question: How close can we come to the visionary fire? And the answer: As close as *myth, magic,* and *mystery* will bring us. These are the primary cultural channels the fire runs along on its way from the Few to the many. Taken together, they comprise what I will call *the sacred triangle,* a three-cornered ground which is the traditional province of religious culture.

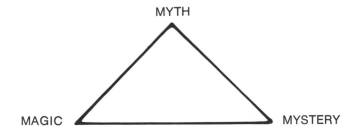

One no sooner writes the words than we have a terrible measure of how degraded the spiritual life of modern Western society has become. Granted, ever since the Romantic movement, there has been a stubborn fringe of artists and thinkers who have insisted on the cultural centrality of myth, magic, and mystery; but what has the popular understanding learned to make of this triad? *Myth?* It is a synonym for falsehood, a species of old literature akin to children's fairy tales. *Magic?* It has become show business hocus-pocus, sleight of hand deception, tricks done with mirrors. *Mystery?* It has descended to the level of detective fiction: at best, a puzzle to be solved, at worst, a dirty secret to be exposed. Nothing of the original dignity of these three concepts penetrates the dense veil of trivialization that now screens them from a decent appreciation in the popular mind.

Instead, as they have declined in popular regard, their high place has been taken over in our culture by more "enlightened" substitutes. In place of myth, we have *history*. In place of magic, *technology*. In place of mystery, *reason*. Here then, we have a second, inverted triangle—a *profane triangle* whose orientation is toward the Earth and away from transcendent experience.

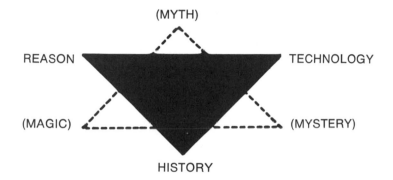

Perhaps this depiction by way of topsy-turvy triangles recalls the famous boast Marx once made—that he was out to stand Hegel on his head, and so to up-end the pipe dreams of idealist philosophy with the iron lever of materialistic science. That puts the matter bluntly; and yet the transformation we are after is just that blunt and bold: one Reality Principle knocking its predecessor for a loop. Late in the course of every great cultural transition, there comes such a moment of sudden, dizzy turnabout, when the new takes over the stage from the old. Within a few crucial generations, a sweeping revolution of values may then shoot through society like pent-up energy that has reached its critical quantum charge. In our case, the great reversal has been the total *secularization* of culture in mind and deed—certainly the most potent, daring, and original project of modern times, as well as the most distinctive historical contribution of Western society.

Yet the very project of inverting the cultural triangle presupposes a sacred original which has, at the very least, staked out some significant space in people's lives, a space we feel must be filled and accounted for. Whatever then occupies this cultural vacuum is bound to inherit something of the power belonging to what came before. The profane triangle is rather like a modern monument that takes over the ground once held by a temple, and before that, since time immemorial, by a sacred grove: it inherits the aura of that holy ground.

That is why history, technology, and reason grip our convictions and aspirations as they do: because they replace the sacred triangle and, by that very act, promise the same fulfillment that myth, magic, and mystery offered before them. And what choice have the secular humanists who worked so hard for this substitution but to insist that the profane triangle, having legitimately overturned the sacred triangle, can and will make good that promise? For them, the two triangles are an absolute dichotomy; the difference between them is the difference between objective and subjective, between wishful fantasies and practical realities, between misdirected psychic needs and rational gratifications. And if they insist that the city of God must make way for the city of man, it is with every honest intention of building a finer, more fulfilling world.

I will not deny to secular humanism its heroic stature. But both by experience and by conviction, I am carried away from its project and toward the primacy of visionary experience. In the two triangles I see one continuous, unfolding reality; but the major purpose of that unfolding is to reveal and celebrate the richness of the sacred source from which profane culture derives. In this view, the primary role of culture is to guide the mind through the profane derivatives that so thickly surround us back to their original visionary impulse; through the million variations back to the theme by which all were generated, until at last we grasp the unfolding of culture as a continuity in which there is no longer a "sacred" that is separate from the "profane," but only so many brighter and dimmer expressions of the sacred. Perhaps then we will see that, like the Sumerian culture hero Marduk, who fashioned heaven and Earth from the shattered remnants of Tiamat the divine mother, we too have all the while been making our world from the body of the god.

Let us see how this happens with respect to each term of the triangles, and what the result is when we invert the proper orders of reality. The movement we are following (and it is nothing less than the course of the Western cultural experience) is from myth to history, from magic to technology, from mystery to reason. The insight of Blake guides us along our way:

The Visions of Eternity, by reason of narrowed perception,
Are become weak Visions of Time & Space,
    fix'd into furrows of death.

From Myth to History

Myth governs the encounter of the person with time, hence with mortality. It uses the action of time—in the form of narrative—to speak of eternal things. In this way, it reminds us that the adventures we experience in time can be, in each moment of their happening, reflections of a timeless meaning.

Where myth touches our lives, we come to feel ourselves actors in a drama which has been improvised upon perennial themes. This is the sense of life that the Greek drama was created to project: a shifting foreground of plot and action, played out against a controlling pattern of mythical motifs. In "real life" (meaning the great theater of the world) the foreground is what we call history, the flux of immediate incident, wild, erratic, confounding. We give this storm of events a meaning we can live by only when we assimilate our foreground experience to the mystical themes which irresistibly reach out to govern our thought and action. The very purpose of myth is to provide the randomness of events with instructive order, so that intellect is not abandoned to the chaos of time. In this way, myth makes history bearable by purging time of that brutal unruliness which leaves nothing certain but transience and death.

But now what happens when the foreground crowds its controlling mythical backdrop out of awareness and becomes the single perceptible reality on the scene? Then history becomes a boneyard of broken faith and perished certainty where nothing has existential structure or moral shape, and each person's autobiography within the flux becomes an absurd struggle against all-conquering impermanence. With all mythical reference lost, a riot of events takes over the stage, each clamoring for equal attention within its "period" or "field." Everything becomes equally worth knowing, and equally meaningless. Historical study turns into the professional task of knitting a million data points together into limited, localized "interpretations" that are seldom more than assertions of the conventional, academic wisdom of the day. Since history is no longer experienced as having the epic stature in which poets deal, scholarship replaces poetry as the diction of history. Footnoted precision evicts rhetorical splendor. So, progressively, history grows richer in fact, but poorer in meaning—except among ideologues and propagandists whose interest is not in mythical paradigms but in useful lies. And in reaction to such contemptible special pleading, principled historians rapidly reach an extreme of learned skepticism, regarding it as backward or plain dishonest to suggest that the past has anything to teach us whatever. What, then, is the past? A formless flow of events at best grouped into specialized fields of study as objectively neutral as the study of chemical reactions in a test tube . . . one damn thing after another . . . a province of research, but not of philosophical instruction.

Myth is therefore both the archetype and the antithesis of history. As archetype, it deals with tales out of time, facts of the spirit to which history may bear witness, but only provided we can see through the foreground of events and recognize the patterns of myth in action—as vividly as James Joyce could see the *Odyssey* being played out for the millionth time in the streets of twentieth-century Dublin. Where we lose this power of *per-ception,* of seeing through, myth becomes antithetical to history, as form is antithetical to chaos, as meaning is antithetical to nihilism. That is not, however, the antithesis which conventional intellect will see; it will instead see myth as "fiction" and history as "fact." And from this conventionally enlightened viewpoint, the depreciation of myth will be regarded as a victory for truth and reason, a welcome enhancement of our ability to see things as they *really* are.

Yet the repression of myth is doomed to self-defeat. Myths are what we have been given to understand our personal and collective experience with. They educate us to the dramatic and didactic structure of history; they offer the paradigms we need in order to read meaning into human conduct. So they cling tenaciously to the mind whenever it addresses itself to history. At most, what the degradation of myth achieves is only a terrible coarsening of mythical awareness, so that the myths come through to us as blurred and distorted fragments of themselves—like images caught in a succession of splintered mirrors. Even so, these twisted caricatures will be the patterns people strive to impress upon the flux of history and to invest with impassioned meaning. At that point, it becomes all too possible for fools and scoundrels to turn the visionary power of myth into the stuff of petty historical ambition. A Hitler can posture before the world as the modern Siegfried, pretending—and perhaps believing himself to be—a guileless child of the primeval forests . . . while leading a regimented and militarized mass movement. In such a grotesque parody, Siegfried's heroic adventure no longer stands as a symbol of transcendence; it is not seen as a whole or experienced to its depths. It cannot come alive in the imagination of the individual as a personal quest for higher self-realization. Instead, only crude images of domination, pride, warlikeness are thrown up before the masses. Only these poor scraps . . . but because they derive from a transcendent symbol, they bring with them the power to arouse heady emotion and to lend sanctity to vicious intentions.

Despite the expert skepticism of our best scholarship, it is the shape of myth that people incorrigibly look for and respond to in history, even without realizing that they do so. But deprived of a robust mythical awareness, they do not assimilate into their own living experience the reenactments of Ulysses and Prometheus, Achilles and Sir Galahad, they see before them. Rather, they crudely reify these images in leaders and political projects, too often using myth in its most debased form to legitimize racist, sexist, or nationalist hostility.

If not good myths, then bad ones. But we will never save people from bad myths by further discrediting the human need to mythologize history. Our task is to educate our eye for myth, not to darken its vision still more.

From Magic to Technology

Magic governs the encounter of the person with nature. Its rituals round about the world are infinite in their variety, but all come down at last to the vision of nature as a live and personal presence. At the root of the magical sensibility is the conviction that mind permeates the universe, filling every natural object with a living, intentional being we may address by prayer, rite, or rhapsodic celebration. So magicians and those who share in their rituals speak to the divinity or the demon they perceive in the rock, the water, the wind, asking for care and instruction, attending at all times to the world about them with that personal rapport we have come to call "superstitious" or "idolatrous." While I am giving the phrase a more unorthodox turn than Martin Buber would have approved, his "I-Thou" relationship is the essential quality of the magical vision of nature.

To one degree or another, this sense of mindfulness in nature lingers on in every religious tradition, if only as a cooling ember left over from the fire of an original animistic vision. But it is weakest in those "higher" religions (especially the mainstream of Judaism, Christianity, and Islam) in which the divine has been elevated above nature, if not wholly divorced from it. It shines through far stronger in the religions we call "primitive" or pagan. There the natural world is experienced to be universally participated by divinity, either by many gods or by an all-pervading, all-encompassing sacred substance. Somewhere between the higher and the primitive religions, we find a number of mystical and occult traditions—Taoism, Tantrism, Kabbalism, Neoplatonism, alchemy—which have managed to provide primordial sacramentalism with sophisticated philosophical credentials and so to bring mankind's most archaic religious experience within the province of civilized society. In effect, these traditions take magic out of the tribal countryside and bring it into the city. Thus, in the Western world, Hermeticism and the alchemical rituals that accompanied it made up the last widely respected school of nature mysticism before the study of nature was monopolized by modern science. As oddly "medieval" as Hermeticism may now seem, one can clearly see behind its quaint lore a sensibility that is shared with American Indian medicine men, Eskimo shamans, Celtic druids. For what is the alchemist's *anima mundi* but the mana or wakan of the world's most primitive tribes: the pervasive divinity in elemental nature?

Since magic frequently takes the form of a petition for favor or a rite directed toward influencing some natural force, it is easy to mistake its aim as power. And, indeed, there is a corrupted sorcery that does seek to force nature to do one's bidding. What authentic magic seeks, however, is not power, but security—the security of being at one with nature, of moving receptively with its grain and sharing its purposes. What is it that true magicians are after? A state of being, not a method of manipulation. In the words of the Rosicrucian oath, the magus "desires to know in order that he may serve." The goal is that awareness which permits us (in Wordsworth's phrase) to "see into the heart of things"— and this we are more likely to achieve by the power of trance than by the power of tools. Another word for security in this sense is, I suppose, wisdom: the wisdom of trust and acceptance that comes of knowing, with the Indian yogi, that *"tat tvam asi"*—"that's you. . . . It's *all* you."

If this kind of security is any sort of power at all, we might call it *being-power*, as distinct from *doing-power*. There is, however, a subtle interplay which often allows being-power to lap over into doing-power, and so to become "practical" in the manipulative sense. So it is that a great deal of technology traces back to magical origins. The agricultural revolution of the Neolithic period was the outgrowth of a magical vision of nature which found its primary expression in the rich poetic symbolism of fertility rituals. So too the early arts and crafts of the human race arose within a context of ritual magic, which made prayer, incantation, sacrifice, and symbolism an inseparable part of all tools and their use. The magical vision hardly leaves people defenseless and abject in the face of a supposedly hostile nature. Rather, it mingles technics with what we would now recognize to be an intelligent ecological restraint. Behind all technology stands the magical bond between people and nature, the first human effort to know, identify with, and wield the great forces of the environment.

But besides the security that comes of trust and cooperation, there is also the security that comes—or seems to come —of domination, provided our ability to dominate can be made absolute. This is the image of security that seduces us into wanting another sort of power—*forcing-power*. Here again the demonic has a shrewd trick to play with us. It convinces us that more security can be gained sooner through force than trust. Especially when we become afraid in crisis, grow rigid and lose our adaptability, the appeal of forcing-power can become irresistible. Once our connection with nature ceases to be a respectful relationship between person and person and becomes the relationship of human master to alien thing, then we have a very different kind of magic: *black magic,* the magic of evil sorcerers who have no wish to *know* the nature of things; they only *use* it for selfish advantage. This was the magic of the "vulgar" alchemists who practiced their rituals to gain the literal and not the symbolic gold of the philosopher's stone: Earthly profit and not spiritual renewal.

Clearly, it is along these unholy lines that industrial society is making its way, driven by its technological obsession. Like Dr. Faustus, we have moved through necromancy toward the diabolical in search of a brute forcing-power over things and people. And just as black magic breeds upon belligerence, seeking to oppress, harm, or exploit, so too our technology has only been able to reach the zenith of its genius as an adjunct of war making. Consider how many of our most astonishing industrial achievements are the spin-off of war, the arms race, the balance of terror—as if nothing except those debased motivations could release our full technological talent.

Again, the path of our culture has been down and away from the source of its inspiration. But even in the latest perversions of our power mania, the magical vision lingers on in distorted form. We still look to sorcerers for our security: sorcerers of the computer printout and the economic index. The esoteric jargon of our bomb physicists and technocratic elites is an echo of the mumbo-jumbo of black magic: incantations of benighted power promising tribal vengeance, dominion, opulence. Only very recently, in the warnings of concerned ecologists, have we been reminded, at least in some small degree, of another security—the security of harmonious partnership with nature. It is a glimmer of the good old magic.

From Mystery to Reason

The mysteries govern the second birth of the person, the encounter with spiritual crisis. Their purpose is to teach our essential identity, the Self within the self, whose discovery may be the end of one life and the beginning of another.

That is normally the situation in which men and women come to the mysteries—at a point when the identity the world has assigned them wears thin or grows burdensome with bad faith. They see the death's head too clearly outlined behind their face; more and more they sense how neurotically brittle is the apology they are forever making for their existence. At last there comes a time when the need for renewal grows desperately urgent within them. They long to throw off the past, expunge its lies and failures, and make a second beginning at a level that transcends self-contradiction. What are they struggling with? Formidable foes—the artificiality of the social ego and the horror of its mortality. Such a crisis of life may lead to breakdown or ugly distortion of the self. But it may also be filled with what Lewis Mumford has called "the optimism of pathology." It is to rescue new life from the death grip of this psychic pathology that the mysteries were devised. They come down to us as the highest and riskiest of all human rites of passage.

Traditionally, as in the mystery cults of the ancient world, one approached the mysteries by way of initiation: a ritualized curriculum of ordeal, instruction, self-examination, and visionary exercise under the guidance of an experienced spiritual master. It was actually these rites of initiation that made the mysteries "mysterious," which meant, simply, unintelligible to the unprepared. Perhaps—again to become purely speculative—the original pattern of initiation was invented by the Few out of their own spiritual experience. They too had arrived at a crossroads in life where withdrawal, isolation, discipline, and ordeal were necessary if the vision that beckoned them forward was to emerge in all its clarity. Perhaps the techniques of initiation appeared as an approximation or reenactment of their life crisis which might now be endured and resolved by others with the aid of their experienced guidance.

In radical contrast to our degraded modern use of the word "mystery," the classic religious mysteries require no deliberate concealment. They may, at a critical moment in the initiatory discipline, require contemplative privacy, but otherwise they guard themselves by virtue of the fact that they can be meaningfully approached only through the course of initiation. How can they possibly be stolen or abused by those who do not know where their value lies? The mysteries may therefore stand before the world as an "open secret" and lose none of their sanctity. This is why the rites and holy things of the old mystery cults were never in any sense "classified" information; they did not have to be. In some of the old cults, the final object revealed to the initiates as the end of their training was known to be something quite ordinary, perhaps only a stalk of grain. But presented in just such a way, at just such a moment in the initiate's experience, when the mood was exactly right, and after all the appropriate disciplines of introspection and moral purification had been mastered, that commonplace image was all and everything. Similarly, Zen Buddhist masters often manage to use the most nondescript objects to convey satori—provided the moment is pregnant and the student ready. Such mysteries are *naturally* unavailable to the hasty and the superficial. As Blake observed: "A fool sees not the same tree that a wise man sees." There is, in fact, no better way to tell where mystery leaves off and exploitative or frivolous mystification takes over: one must work hard at protecting dirty secrets and deliberate obscurities—as hard as secret police must work to ensure the security of the "mysteries of state" they exist to guard. True mysteries, on the other hand, are hidden only by the spiritual blindness of the uninitiated.

This is the main reason why the knowledge of mysteries is always said to be "beyond words." Not because that knowledge cannot be communicated, but because the mode of communication it requires is an initiatory process which leads to seeing with one's own eyes. Spiritual masters are notorious for their elusiveness; they simply will not pass the mysteries along in the form of reports. Rather, their way is to *arrange* for the mysteries to be learned by direct, personal understanding; they create spiritual environments in which the desired experience may flower. What did the Buddha offer his disciples? Not a creed to be memorized, but an eightfold path to be followed; a path that required nothing less than a radical transformation of life in body, mind, and deed. And beyond that path lay only the plain of knowing silence.

In the modern West, popular familiarity with spiritual initiation degenerated catastrophically in the wake of the Enlightenment. Remnants of initiatory discipline lingered on in secret societies like Freemasonry and the Rosicrucians, and in the novitiates of some religious orders, but all these grew less and less visible to the general culture. It is not until the late nineteenth century that we find a scattering of marginal occult groups—like the order of the Golden Dawn, the Theosophists, Rudolf Steiner's Anthroposophists, and the disciples of Gurdjieff—making any deliberate effort to resurrect lost rituals of intiation or to invent new ones. In recent years, of course, these Western improvisations have been joined by a variety of Oriental initiation lore, imported mainly in the form of meditation techniques currently taught by a small army of gurus, swamis, lamas, and yogis. These, in turn, have flowed into the more adventurous kinds of Western psychiatry to create the new eupsychian therapies which more and more come to look like psychologized versions of old initiation rites. One weekend workshop in transpersonal therapy at Esalen Institute, for example, offers a "tender invitation"

to go beyond where you are through seeing how you keep yourself behind. Beyond is a rich alive experience—love, anger, joy, sadness and spirituality. Behind is a fear of seeing what is there. The participants will be offered a chance, a tender invitation to see what is behind and go beyond. Traditional encounter, gestalt, bioenergetic and psychosynthesis techniques will be utilized in addition to a new approach aimed at developing perspective, the ability to "see oneself beyond the impasse."

Such experiments are at least flirting with the emotional needs that once brought people to the traditional rites of religious passage, though how far a weekend of eclectic therapies will carry its participants is another question.

Despite the efforts we see around us to reclaim a vanished spiritual discipline, only sadly debased strains of religious initiation have survived in the public awareness, mainly in social clubs and fraternities, where inconsequential secrets are passed on to new members, while made-up rituals are enacted and meaningless harassments endured. Surely not many who undergo such nonsense rites can realize how old a tradition these shallow exercises belong to, or what awesome adventures of the spirit they blithely mimic. Beyond this, the only prominent examples we find of initiatory discipline in the Western world have been thoroughly secularized, often with the bleakest results. In its most degraded form, initiation has been harnessed into the brainwashing methods of totalitarian political movements whose only values are vengeance and power. Here, with no transcendent purpose in view, the object is simply to "liberate" distraught initiates from their personal anxieties by feeding them body and soul to all-devouring party leaders. This is to relieve the tormented ego at the expense of murdering the spirit.

We can also glimpse forms of secularized initiation in the way people are prepared for lifelong service in some professions or in the major corporations. Young initiates entering these careers may pass through a rigorously systematized curriculum meant to ingrain the worldview of the organization or to fit their consciousness to a prescribed reality. In science and medicine, for example, much of the student's education—especially the aura and ritual of laboratory, clinic, dissection room, or the conventional style of textbooks and published research—is designed to shape the sensibilities as much as to inform the intellect. So too the subtle, personal pressures that teacher exerts upon student amid the fierce, competitive tensions of graduate study: the disapproving tone, the raised eyebrow, and the censorious glance that are sure to greet any remark that strays toward Lamarckian biology, Velikovsky's astronomy, or nonrelativistic physics. The object of such subliminal pedagogy is really to maneuver students into a way of seeing—and of *not* seeing—the world around them.

Yet, oddly enough, when we turn to the religious mysteries, the importance of emotional and psychic preparation of
a far more deep-reaching and discriminating kind is often
summarily dismissed by both ardent believers and skeptics
alike. In many mainstream churches, the deep truths of the
faith are apparently meant to be learned from dry dogmatic
instruction or rote catechism wholly without the aid of
visionary insight. In the pentecostal congregations and
among many countercultural occultists, on the other hand,
every psychic tremor that sweeps across the most untutored
mind is immediately endorsed as a divine visitation. Meanwhile, militant skeptics will insist on having the meaning of
mysteries fully and cogently revealed on demand for the
inspection of wholly uninitiated or even hostile eyes. In all
these cases, the necessity of careful psychic preparation and
a mature power of discrimination is simply unrecognized;
either anything goes or nothing goes.

There is an elitist ring to the idea of initiation, and it is
understandable that this should be off-putting to democratic spirits. What is overlooked, however, is that the democracy of the mysteries lies in the access granted to initiation, not in the publicity of what initiation teaches. The
mysteries guard themselves—and about that we have no
choice. Still, the path that leads to them can be open to all,
one by one by one. But those who demand to know, before
they have set foot upon the path, what will be found at the
end are fools, and the duty of the honest guru is to say so,
and perhaps to break off instruction until the need for rebirth has intensified.

Now what happens in a culture like ours where the mysteries cease to be an experienced insight and harden into verbal
formulations and routine gestures? Then initiation degenerates into mere catechism, and the traditional forms of
spiritual instruction—the images, symbols, catch phrases that
were so many thresholds to visionary experience—become an

opaque medium. At that point, simple mystagoguery is not far off. Sensing that the mysteries are not meant for untutored eyes, their latter-day and lesser guardians begin to erect barriers around the words, objects, and exercises that have been used to convey them. It is these "visual aids," formulas, and techniques of instruction that are now taken to be sacred. The natural secrecy that surrounds true mystery is replaced by deliberate concealment. The shamans and gurus who lived to share their vision by way of initiation are gradually crowded out by priestly elites who mistake the essence of religion to be premeditated obfuscation. Soon the sacred becomes too holy for any to touch; it becomes the totally unknowable, the absolutely forbidden. Its various formulations, lacking the vitality of experience, become sheer mumbo-jumbo. Obscurantists rush forward to thwart any approach to the mysteries—or, worse still, to charge admission for a glimpse of some inferior icon of the cult. The natural (but purely provisional) hierarchy between master and student is debased into caste privilege, and anyone who hungers for more than the priesthood has to offer is outlawed for heresy or sacrilege.

So, by this downward spiral, we arrive at that caricature of religion against which, in our society, the best minds of the Enlightenment rose up in justified wrath. But what was it that the anticlerical skeptics and crusading agnostics recommended as a remedy for the corruptions of the established churches? Not the reclamation of the mysteries or the renewal of initiation; the memory of this had been too much dimmed. By the time we reach the era of Voltaire, the practice of religious initiation in the West was all but dead; the few surviving occult traditions had been driven underground into secret societies; the gurus and spiritual masters were nearly an extinct breed. At a critical moment in our history, those who knew or instinctively sensed the importance of mystery in the life of the mind—those like the Freemasons, William Law, Swedenborg, Thomas Taylor, William Blake—found themselves without sufficient audience among the intellectual vanguard and the aroused bourgeoisie. The indignant tide of Enlightenment swept over them, demanding a new standard of intellect. "Reason," it was called, Reason-with-a-capital-R. And it meant full and immediate public access to all knowledge, on the model of the new natural sciences. Knowledge was now to be open to all comers, fully articulated, rigorously logical, instantaneously self-evident, conveniently empirical. In short, knowledge had to be either obvious fact, logical deduction, or common sense. But of course, the mysteries cannot be embraced within so narrow a program. There is no way into them except by the path of initiation. *In this sense,* they cannot be thrust into the public domain, even though the rites of initiation remain accessible to everyone.

Our society has rejected the mysteries with all the best intentions. We have cast them out in pursuit of reason and freedom. There is both glory and tragedy in this episode: the glory of great ethical idealism, the tragedy of spiritual blindness elevated to the status of high principle. Yet, as in the case of myth and of magic, here too the visionary impulse survives and continues to act upon our lives, though, again, in a subterranean and distorted fashion. By subordinating the mysteries to militant rationality, we have saddled reason with the function of gnosis, which is to ask the impossible: that the part play the role of the whole.

We call the virtue "reason." The name suggests a cool, cerebral quality of mind: logical, restrained, judicious. But not far below the surface, reason reveals itself to be a towering moral passion, an energy that has launched great revolutions and in whose name people have killed and willingly been killed. A passion for what? For free and equal access to a knowledge that we sense our entire dignity demands. Convinced, as so many of our keenest intellects have been, that this knowledge must be the sworn foe of religion, we look for it everywhere but in religious tradition. We become the most knowledge devouring culture in human history, obsessive collectors of fact and data; we research all the world as if our very salvation depended on the result. But where, in the first place, did such a hunger of the mind come from? How did we become so convinced of the ultimate value of knowledge?

If only we would search this passion called reason to its roots, we might be startled by what we found. It is gnosis we are still pursuing—a knowledge of the mysteries acquired on the threshold of ecstasy. When will reason, the lost, rebellious child of that visionary energy, find the wings that belong to its parentage?

Blind Alley

At the outset of this chapter, I observed that the profane triangle is an inversion of the sacred triangle. And that is indeed how the matter has appeared for the past two centuries to many crusading secular humanists in their urgent quest for revolutionary solutions. From that viewpoint, the task of secularization has been to diminish, if not destroy, the place of myth, magic, and mystery in the life of the mind. Regarding the two triangles as antithetical, secular humanism has insisted that a mature, scientifically sound culture is the flip-flop opposite of religious tradition. At most, it pluralistically tolerates the surviving remnants of the sacred triangle, granting them a tiny, diminishing enclave within the dominant culture.

But what is the result when secular culture so massively engulfs the sacred triangle? As we have seen in this chapter, the outcome must be traced along two lines: that of the intellectual mainstream and that of the general public.

For the intellectual leadership of our society, each term of the profane triangle was originally seen as a means of building the heavenly city here on Earth as a "real" social project. But, in the course of time, each has proved less and less capable of yielding the shining human benefits it had promised. Two centuries ago, when the best minds of European society undertook to replace myth with history, they saw that exchange as the precondition of limitless secular progress. Today, there are few thoughtful heads who have not long since learned to identify progress itself as a "myth" in the most pejorative sense of the word—a pipe dream. So the hope of giving mankind an ennobling but wholly historical identity has been lost in a wild turnover of social ideals and ideological crusades none of which seems to bear any values that outlive its generation, all of which finish in so compromised or flawed a condition that they serve only to enforce bitter disillusion. At last, disillusion itself becomes a crusading cause that attracts to its grim service growing numbers of hard-boiled intellectuals whose whole mission in life it is to use some current fashion in "realism," "pragmatism," cynicism, or outright nihilism like a meat ax on the public's naïve faith in progress. If Western society's waning confidence in worldly progress—especially in the decades since World War I—proves anything, it is that history devoid of myth can only dissolve human experience into a relativistic fog in which the mind finds no universal meaning, unless it is the lesson of world-weary resignation. *Vanity of vanities* . . .

As for technology, which was meant to serve as the engine of progress, it becomes a compulsive and self-defeating pursuit of total dominance over society and nature. But, as we now learn more vividly from each new economic and social breakdown in our system, under the influence of so impossible and unbecoming an ambition, urban-industrial society is doomed to lurch from crisis to crisis, from emergency to emergency. And who cannot already foresee the final act of this unhappy scenario? A global wasteland where only an ensconced, bandit elite of corporate profiteers, commissars, and technocrats, together with their armed defenders and reliable clients, enjoy the dwindling riches of the Earth, while the impoverished billions starve without even clean air in which to draw their last breath.

Finally, reason, in its principled search for the truth, turns into an annihilating skepticism which can deftly dissect illogic and superstition, but which gains its analytical prowess at the cost of driving ever more of our transcendent aspirations into a locked and barred mental ghetto of "irrational" impulses. Once, in the bright sunrise of the Enlightenment, when it was confronted with the persecuting obscurantism of a corrupted religious establishment, reason— in the form of a brave and admirably clear-sighted skepticism —performed a great humanitarian labor in our culture. But now, with so much of its original clerical opposition crushed, reason reveals itself as, essentially, a negative talent of the mind that shines only in opposition; otherwise, it has no life-giving destiny. So, it finishes as mere functional rationality, capable of producing abundant research and technical know-how, but not even the shadow of a sustaining wisdom.

Meanwhile, the masses of our society, starved of the spiritual nourishment that the sacred triangle exists to supply and cheated of the progress that secular culture has promised, are left to make do with trivial or demonic substitutes for authentic myth, magic, and mystery. So we follow our course down and down into doubt, confusion, and despair. It is as if we have pursued the sun's inviting reflection upon water down through the surface, deeper and deeper into the murky depths of the sea. There, the traces of sunlight that reach us are only elusive glimmers playing here and there off scattered debris. But having come so far, what can we do but rush furiously about to collect the debris, insisting to ourselves that *this* is the source of the light we have been after, that, at last, we have got to the bottom of the matter.

The figure of speech I use here—that of the sunlight and its watery reflection—is carefully chosen. We have already touched upon the classic teaching in occult traditions like Hermeticism and alchemy which holds that the world below mirrors the world above, and to that degree is charged with the splendor of heaven, serving as the symbol and epiphany of the transcendent. But, the teaching goes on to say, it is essential that the mirror be recognized *as a mirror* if it is to direct the mind to the source of its reflections. If that fails to happen, there ensues a fearful, indeed a *demonic* confusion. We begin to believe that the reflection *is* the reality, that there is no other reality to which attention is due, that our visionary energies are therefore superfluous, infantile, misguided, demented. Once believe that, and the orders of reality are inverted. The shadow becomes master of its substance, the reflection blocks out its original, and we have no place left to seek fulfillment except among shadows and reflections.

That condition of benightedness goes by many names. It is the Buddhist samsara, the Hindu maya, the Platonic cave of shadows. It is the fallenness of Christian experience, the alchemist's *nigredo*, St. Paul's "carnal mind," Blake's kingdom of Ulro. All these are so many names for the world of everyday life when that world can no longer fire the mind with transcendent awareness. At that point, it is the habit of those who guard the vision of the Few to do a strange and unnerving thing. They disparage *this* world in the name of another, *higher* world, a world they call more *real* than this.

But the "otherworldliness" of the great religions need not be a rejection of this world's reality and magnificence. It need not be read as the statement of an irreconcilable dualism—though it is frequently mistaken for that. Rather, the teaching counsels us to see that the unity of "this" world and "other" world is the relationship of a reflection to its source. The reflection is also a reality to be appreciated and enjoyed, all the more so when we grasp the role it plays as the image-bearer of a transcendent original. Only when it fails to perform that role—or rather, when we fail to use that potentiality within culture—does it become an illusion, a mirage, a veil.

Religious teachers who have, in running to compensatory extremes, lost sight of the evolutionary unity which makes the profane triangle a wholly natural transformation of the sacred have cast away the one truth that makes them competent to minister to our whole human nature. But at least that truth remains part of their heritage, if it is not indeed the fountainhead of our religious sensibilities; it is there to be found in the symbols and mysteries of the spiritual life if they will only search far enough. On the other hand, the secular humanists, who have risen in rebellion against the vices of established religion, have, within their brief tradition, no such truth to rediscover. Their commitment is grounded in an angry alienation from sacred culture. As a matter of principle, they have cut themselves loose from transcendent experience, confining their world view to the perimeters of the secular triangle. If they would become ministers to our whole nature, and not only to our vanity and moral outrage, they will have to enter their longtime enemy's camp to find instruction. They will have to make themselves students of the sacred traditions they have come to regard as empty and out-moded. Indeed, they will have to reach deeper into those traditions than the mainstream clergy have done in Western society—and at last deep enough to find the visionary energies from which the great human adventure in culture-making takes its course.

Which is what we see happening already along the Aquarian frontier, here and there, by fits and starts, as small scouting parties of artists and scientists, scholars and therapists, free themselves from the bad reductionist habits of the past and turn to the swamis and seers to learn the proper place of myth, magic, and mystery in our lives. It is the risky birth of a new humanism, a visionary rather than a secular humanism. Certainly if humanism has any life-enhancing future in modern culture, it lies in granting to each of the triangles its proper function: to the secular triangle the role of celebrating its sacred original, and to the sacred triangle the role of containing its profane reflection.

8

The Higher Sanity and Its Competitors

Age of Faith . . . Age of Reason . . . Age of Anxiety . . . and how is the age that opens before us to be remembered? Can there be any doubt? The Age of Therapy. When all the ailing souls came home to be healed, and sweet sanity inherited the earth.

But whose sanity? There are many. And behind each there stands the prospect of a different humanity, a different reality. To enter the deep psyche is to enter an intimate battleground where the future of our society is being contested.

This is not a fact that enjoys much visibility in the psychiatric profession—except at the dissenting fringe. Psychiatry is the modern world's way of talking about the deep nature of man as if we were talking medicine, never morality, never metaphysics. So we dissect the heart's unease into symptoms and complexes, taking comfort in the clinical ring of the diagnosis. Perhaps, if we finally associate enough pills and potions with professional treatment, sanity may even come to look like no more than a pharmaceutical concoction.

Even among psychiatry's innovators and mavericks, there are many who go no further than to advertise dazzling forms of emotional engineering. Each year another therapeutic

182

Tom Swift invades the book stalls with the latest rage in popular psychomechanics guaranteed to oil the friction out of our human relations and tune up our egos until, like components of an efficient moon rocket, you and I are also "A-OK." But for all this flashy traffic in psychiatric fashions, the curing of souls remains what it has been since time out of mind: a religious labor, the age-old vocation of priests and shamans. The thin skin of scientized jargon we have stretched over this venerable art scarcely hides the body of implied metaphysical commitment beneath. To prescribe sanity, we must first risk saying what is humanly normal. And to say what is humanly normal, we must know—or assume we know—what the nature of the human animal is. Scratch a therapist, find a philosopher—though perhaps a philosopher in spite of himself.

What I have suggested in the preceding chapters counsels a profoundly conservative approach to therapy: conservative in Paul Goodman's sense that these days nothing is more radical than to be a "neolithic conservative" out to save the forest and the river, the village and the old crafts. So I am concerned to save the craft of therapy by divorcing it from the scientific self-image that makes it so shallow a trade, and returning it to its source in religious culture where it can alone find the metaphysical insight that does justice to our nature. In what sense is the sanity I speak of "higher"? In that it builds its model of normality on the visionary awareness whose origin I have attributed to the Few, encouraging us to see our human identity as a rich potentiality that can be completely unfolded only by an etherealization of life. With such a model before us, we might even come to recognize that unfolding as an evolutionary movement of our species by which we approach a goal transcending time and matter—a goal that, once perceived, enlivens in us that subtle energy of the personality which our ancestors called "spirit."

For the modern mind, this higher sanity is bound to be

an unsettling prospect, not simply because it is bizarre, but even more so because it places us in the humbling position of having to borrow wisdom from the remote and even primitive past—as if we did not know it all and know it better. The higher sanity returns us to the origins of religious consciousness; it makes us the students of ancient sages and sorcerers. It requires that our image of what is humanly normal make generous place for desert prophets, contemplative hermits, ghost-dancing shamans, ecstatic saints: the very types who have for so many generations served as examples of all that is backward and outlandish. Yet is from such unlikely origins that we inherit the symbols which best express our transcendent needs: the jewel in the lotus, the philosopher's stone, the resurrection from the dead, the ladder of the sefiroth, the dance of Shiva. . . . For the higher sanity, these great images —not yesterday's clinical research or tomorrow's behavioral findings—constitute the true subject matter of human psychology. We must see that Western science, including our psychiatry, has fashioned us no better means to negotiate the transcendent level of experience. These inherited symbols are the diamond that cuts all else: our task is to bring their root meaning back into the circle of our experience, to invent the ways that will make them a transforming power in our lives.

Health means wholeness means holiness. The higher sanity grounds itself in that venerable equation, confident that any medicine which works from lesser principles can provide no true therapy. Still, it takes the field against many determined competitors. In the Age of Therapy, psychiatric schools multiply among us like the religious sects of the Reformation— and the relations among them are often as bitter as the relations between orthodox and heretical confessions. From the viewpoint of the higher sanity, each of its opponents touches upon some aspect of psychological truth that will have to be preserved by any adequate therapy. But in each case, the notion of sanity that is offered is too narrow to let the personality grow to its full natural size. What follows is a review of the major conceptions of sanity which modern psychotherapy has developed. Let us see what the higher sanity can learn from its competitors, and at what point it must part company with each.

Freud: The Sanity of Tragic Humanism

All psychiatry is a search for the hidden self. And this, as Freud fully recognized in devising his standard of sanity, is the traditional concern of religion. He more than once recalled the introspective task which the god Apollo assigned to Socrates and to which the philosopher was summoned again and again by his mystic voices: "Know thyself." But for Freud, the consummate nineteenth-century positivist, self-knowledge meant the knowledge of personal causes, not transcendent needs, of organic appetites, not spiritual purpose. Freud's conviction was that the quest for the self must take us down and back—into the juice and tissue of our physical nature, into its infantile fantasies and passions. The way to sanity lay through the history of the body and its many thwarted gratifications.

For Freud, as for Marx, the struggle of the age was against "false consciousness," the substitution of fantasy for fact. Marx saw culture as so many "phantoms of the mind" whose lofty pretensions only veiled the bedrock reality of class war and economic necessity. Freud took his reductionism even further—indeed, to an extreme which no Marxist or Marxist society has been willing to tolerate. Freud's contention was that even the high ideals of political struggle and the compulsions of economic life were part of the cultural masquerade: so many sublimations worn by our tormented eroticism to disguise its wounds and wishes. Mankind's entire history-making energy, including the ideologies of social justice, was the shadow play of instinctual conflict.

Though Freud was convinced that religion was the most infantile of all our cultural fantasies, there is a special irony about his critique of the spiritual life. His intention was to reduce religion to its biological and psychological origins, to disenchant it ruthlessly, and at last to replace it with a new, wholly secularized Reality Principle. The task led him ever deeper into the personality, to levels of repression that few of his followers have dared to fathom. By the end of his career, as he wrestled with the implications of the death instinct, he had reinvented within psychoanalysis an epic conception of life which forced him to draw upon the language of myth as freely as any prophet. Eros and Thanatos in their dread encounter are little short of cosmic entities, worthy successors to the gnostic forces of light and dark or to Blake's warring Zoas; and the story Freud tells of their interplay is, for all his dour positivism, as visionary a tale as any mystic ever told. In his own tortured, halting way, Freud had hammered out a materialist mythology which would at last allow him to write the wisdom literature toward which his whole life's work had been tending.

The level to which Freud traced psychic conflict in the human mind grants a *seriousness* to the problem of repression which is unmatched in modern psychiatry. No one has presented therapy with a more insoluble dilemma than Freud in depicting the confrontation of the life and death instincts. There is a brooding and fearful drama in that dark antagonism which the higher sanity must surely appropriate if it is to avoid the smug and superficial nostrums with which too many new therapies abound. Indeed, so irreconcilable did Freud make the antagonism of Eros and Thanatos that his drama could only have a tragic resolution. Ultimately, his therapy could only counsel a wise resignation to honest despair. "Acquiescence in fate" was what Freud grimly recommended, convinced that the personality dead-ends at its instinctual basis somewhere within the mute realm of the inorganic. Rather than seeing life and mind rising out of the inorganic as the natural and triumphant destiny of matter, Freud, yielding to the pessimism of the newly dis-covered law of entropy, could only envision life descending hopelessly into cosmic dissipation. Like many another un-flinching materialist of his time, he found an almost godlike absolutism in matter that was not to be questioned—even after the physics of the modern world had itself lost its sure grip on material nature and had begun to perceive, at the bottom of things, a range of bewildering and spectral energies whose full potentialities may remain forever enig-matic. He could not see that, instead of finding some hard, cold, and dead foundation of "purely physical causes" be-neath the world, science had dug its way out through the bottom of a floating island that hangs miraculously and mysteriously unsupported in the void. The religion of mat-ter has proved to be among the most short-lived faiths in history.

For Freud, sanity meant the surrender of all illusions and the final acceptance of our total isolation in an alien universe. It is a vision of the human condition not unlike the "dark night of the soul" in which the mystics find their most an-nihilating experience of cosmic abandonment. There is an undeniable Stoic heroism about such a willingness to face the dreadful: a courage and sobriety that puts to shame the frivolous superficiality which other psychiatries bring to their investigation of human nature. But what escaped Freud wholly was the possibility that this terrible state of being, this extremity of alienation, might itself be the greatest of all illusions: not the cure our wounded psyche requires, but the disease in its most concentrated and toxic form. But so the higher sanity takes it to be, following the lead of those who have crossed this desolate territory and fought their way free of its despair.

Behaviorism: The Sanity of Adjustment

The goal of behavioral psychology is to shape us to the world-as-it-is by smoothing away all our rough, unsociable edges. Its fixed beacon of normality is the going social consensus, especially as voiced by official authority. Its self-appointed task is to serve that consensus, not to question it. Therapy, for the psychic adjusters, is the fine art of lubricating disaffiliated personalities back into productive roles within the social system. As for those whose deviance sticks hard, the adjusters have in recent years developed an array of formidable techniques: aversion and behavioral therapy, operant conditioning, minutely calibrated "schedules of reinforcement," electrical stimulation of the brain, pharmacological and even surgical procedures for altering personality —all together, an arsenal of manipulative weapons that makes "adjustment" indistinguishable from out-and-out brainwashing. At the extreme, the adjusters claim the power to redesign anyone to fit any social blueprint simply by the proper application of negative and positive reinforcements—as if we had no basic nature that deserved respect. In their ethical neutrality, they mistake what *can* be done for what *ought* to be done.

Within the last two decades the Sanity of Adjustment has come under heavy critical fire from both lay and professional quarters because of its willingness to accommodate itself to any status quo no matter how diseased. (The Humanistic Psychology of Abraham Maslow and the Radical Therapy of Claude Steiner have been among the most important professional responses; the novels *1984* and *A Clockwork Orange* have made the case most vividly for the general public.) The criticism is well deserved. Works like B. F. Skinner's *Beyond*

Freedom and Dignity, a recent advertisement for the adjustive ideal, are a scandalous betrayal of our natural rights. The "technology of behavior" Skinner advocates grants nothing to the autonomy and originality of the person. It denies the existence within us of a moral nature, and in so doing leaves us vulnerable before social power. This, in the teeth of the fact that we have in the Soviet Union today a terrifying example of where Skinner's psychology leads: to a society where the psychiatric hospital becomes a favored instrument for denigrating and breaking dissent. And the more courageously dissenters resist this therapeutic aggression upon their personal integrity, the more insane they are judged to be.

Moreover, Skinner fails to realize that whatever standard of normality society raises, it can do nothing more than draw on the achievements of creative minorities of the past. All things human have to be invented, even the ideals we abuse and betray. Today's normality is a collage of stereotypes based on the teachings and lifestyles of yesterday's ethical and artistic innovators. If we are prudes in one generation, it is because of what Calvin and Wesley and Jonathan Edwards made of the Christian tradition; if we are permissive swingers the next, it is because of what Freud and D. H. Lawrence, Wilhelm Reich and Henry Miller have taught us about the carnal delights. Because the Sanity of Adjustment cannot accept the troublesome fact of creativity, but in fact seeks to extinguish it, its therapy is, at best, a prescription for stagnation; at worst, the instrument of totalitarian control.

With all that said, however, there is this much we can credit to the psychic adjusters. They know that sanity ultimately demands a social context and public standard. And that must be the ideal of all true therapy: to bring the individual and the community into harmony. Given the gregarious animal we are, this is one of the indispensable aspects of psychic wholeness. But adjustive therapy fails to see that society may be *more* responsible for thwarting that ideal than the individual. It conveniently adopts the social consensus at hand as its standard, as if it were not obvious that, within living memory, most civilized societies have been collectively far crazier than their craziest members. The proverbial lunatic is deemed mad because he play-acts at being Napoleon. But Napoleon himself really *was* Napoleon and an entire society endorsed his megalomania and followed him on his mad, murderous escapades across the face of Europe. In view of such collective insanities (what Lewis Mumford has called "mad rationality" and C. Wright Mills "crackpot realism"), the best we can say in behalf of any historical society is that it may have allowed sufficient dissent to flourish around its edges to repair itself somewhat, or at least to keep alive the hope of improvement.

What is the Sanity of Adjustment, then, but a caricature of the therapeutic ideal? We may grant that the good community owes its members a model of sanity to support their growth and to guide them home when they stray into neurotic byways. But where, in our world today, does such a model reside? The National Security Council? The Board of General Motors? The behavioral psychology laboratories of Harvard's William James hall?

Radical Therapy: The Sanity of Social Relevance

The new "radical therapy" is a direct, principled response to the reactionary politics of adjustive psychiatry. The radical therapists wisely recognize that sanity cannot be consummated outside the healthy community. But they are perceptive enough to know that the healthy community does not yet exist, except as an ideal. So they strive to achieve that community by revolutionary change. Unlike the psychic adjusters, whose allies and principal clientele are the mandarin classes of state and corporation, radical therapy makes common cause with the exploited and excluded: with women and the underdog minorities, with the poor and the socially deviant. It knows how cruelly the psychology of adjustment can be used against these groups as a weapon of social control, just as it knows how effectively establishment psychiatry functions to tranquilize the secret moral panic of its middle-class customers.

Radical therapy is social action with maximum self-analysis along the way. It works from the premise that people cannot save their own souls without actively addressing the evils around them, because those evils are the fundamental cause of neurosis. The "maladjusted" among us are to be seen as the casualties of a vicious social system. Their psychic anguish is not, therefore, to be pacified, but to be *clarified;* it is to be used as a lever for the raising of political consciousness. What, then, is the responsibility of therapy? To cure the system by organizing its victims for militant resistance. For this reason radical therapy prefers to work with groups who share a common experience of exploitation, so that it can help people contact others in the same plight. Claude Steiner's formula is: "liberation = awareness + contact." [1]

The genealogy of radical therapy is closely tied to the "Freudo-Marxism" of Wilhelm Reich, the first of the post-Freudians to emphasize the political context of neurosis and the revolutionary calling of psychotherapy. But there is an important difference. Reich believed that individual therapy—aiming at the achievement of the ideal "genital character"—would of itself snowball into a revolutionary movement against the incorrigibly "fascist" nature of capitalist society; radical therapy insists that therapy and revolutionary politics must be pursued and *organized* simultane-

1. Claude Steiner, "Radical Psychiatry: Principles," in Jerome Agel, ed., *The Radical Therapist* (New York: Ballantine Books, 1973).

ously; indeed, that the core of therapy is political action. Reich's political context has become radical therapy's major psychological content.

The moral courage and humanity of radical therapy are undeniable; no one who comes out of a radical political background can find fault with its insistence that injustice and oppression warp the lives of people in ways that can be far worse than the damage done by, say, an inadequately resolved Oedipal attachment. The trouble is: for all its ethical nobility, radical therapy borrows too much of its energy from old ideological anger. For it, as for the old ideologies, to "raise consciousness" means to *socialize* consciousness—by giving it a dimension of class, sex, or racial indignation. For those who begin in a stupor of bewildered personal frustration, an informed indignation is admittedly an expansion of consciousness. It is the first important step beyond helpless isolation into the world of affairs at whose hands they suffer.

Yet being personal—*purely* personal—is not always the result of social isolation. Nor must it always be regarded as copping out. There is, properly and naturally, a personal dimension of life which we seek out voluntarily and where we find experiences the world of affairs cannot offer. That personal dimension is not antisocial; it is transsocial. It is that within us which transcends the cause of the moment: the life-giving quiet we must always find at the center of the relentless social storm. It is the place where dreams happen and where art finds its sources. (It is significant that radical therapy has no psychology of dreams or of art beyond, perhaps, a social analysis of their contents.) There is a great project waiting for us in that quiet place, a project that requires of us a deep inwardness; and its achievement is as relevant to our sanity as political action.

When Claude Steiner argues that "all alienation is the result of oppression about which the oppressed has been mystified or deceived," he is at most talking about *social* alienation, the exclusive preoccupation of most left-wing politics. But the alienation in which Kierkegaard, Nietzsche, Kafka, Beckett, Camus deal cannot be reduced to class relations—not without trivializing their experience. There is a social level of alienation and a cosmic one. There is an alienation that may be remedied by socializing the means of production; but there is another alienation—a condition of spiritual disconnectedness—that will still be with us after all the expropriators have been expropriated. And there is no guarantee that the condition will be more easily cured then. Indeed, both the Soviet Union ("the workers' fatherland") and the various people's republics are more deeply entrenched in *this* form of alienation than we in the capitalist West; they simply refuse to let it be named and expressed.

In the late sixties, a group of politically engaged psychologists and psychiatric social workers launched a publication called *The Radical Therapist*. Its brief history tells a revealing story about the Sanity of Social Relevance. At the outset, the journal emphasized critical commentary on the treacheries of conventional psychiatry, the political bias of mental health programs and institutions, the special and unmet social-psychological needs of women, blacks, Indians, and the poor generally. The coverage was angled toward the peculiar concerns of therapists; the tone, while radical, was unapologetically professional.

Then, after only some two years of publication, the staff decided that there really was no specifically professional task for therapists to perform in our society. More and more the main issues of the day were political and required a commitment to go among the people for full-time agitation and organization. They argued:

It is clear that therapy cannot provide solutions to the institutions of our society. . . . Revolutionary change requires total restructuring of institutions along with the breaking down of false power divisions within them.

From this realization we make several moves. One is from professional reform to social revolution. Another is to see that psychological oppression is not simply one of many types of oppression, but that it is an all-pervading aspect of our society, of modern capitalism.[2]

So then, therapy could wait; or perhaps it would happen automatically as a by-product of social revolution. Perhaps, after all, Lenin, Mao, and Fidel were the world's first real therapists. Accordingly, *The Radical Therapist* changed its name to *Rough Times*, downplayed its professional emphasis, broadened its political perspective, and became very nearly indistinguishable in tone from any number of standard left-wing publications. Psychology had given way to activist sociology; the therapists had all but dissolved themselves in the social stew.[3]

Did this denouement result from a realistic picture of how urgent the crisis of the time is? In part. But far more so, it resulted from a diminished conception of what therapy ought to be. And that, in turn, grew from a narrow conception of our human needs, which are greater than social responsibility alone can comprehend.

The ethical strength of radical therapy must be included in any valid standard of sanity. Any therapeutic effort that is not willing to stand up to the wrongheaded ruling classes of the Earth and resist their incursions is either frivolous or cowardly. Tick off the psychic ills that most torment the world we live in. Power mania, vindictive paranoia, counter-

2. From the introduction of Jerome Agel, ed., *Rough Times* (New York: Ballantine Books, 1973), p. ix.
3. *Rough Times* has since become *RT: Journal of Radical Therapy*, available from *RT*, Box 89, W. Somerville, Massachusetts, 02144. Also see *Issues in Radical Therapy* (2901 Piedmont Ave., Berkeley, California, 94705) and *Madness Network News* (2150 Market St., San Francisco, California, 94114) for the ongoing discussion of "psychiatric oppression."

phobic masculinity, anal-retentive venality, delusions of omnipotent grandeur . . . where do we find these pathologies more influentially displayed or more fanatically defended than at the level of governing elites? *There* is the psychotic resistance that most needs to be therapeutically challenged and politically broken. Yet, for all its libertarian intentions, radical therapy—precisely because it is so all-consuming and so highly principled in its political orientation—could well achieve the same result as the covert manipulations of the psychic adjusters: the totalitarianization of life. Today, in socialist societies like China and Cuba, "consciousness raising" and the inculcation of class solidarity have become an obsessive preoccupation of the state, as heavy-handedly unrelenting as the catechizing of the faithful in Calvin's Geneva. The object is to drown private life in a sea of social engagement. Such enforced altruism on the part of those whom Steiner has called "lefter than thou" types is not a virtue; it is an atrocity. How does it really differ from Skinner's desire to engineer good citizenship by manipulating the "contingencies of reinforcement"?

R. D. Laing: *The Sanity of Madness*

Although the standard of sanity I mention here is closely related to radical therapy, it deserves separate mention because its leading proponent, R. D. Laing, has given it a special and provocative twist.

Laing's argument is that madness is a social creation that cannot be reduced to strictly intrapersonal causes. It happens *between* people, even, or perhaps very often, between therapist and patient. It is a subtle mixture of people's mutual experience: of their overlapping misperceptions, ploys, stratagems. As such, madness, on the part of the wounded and abused, is a legitimate form of psychic self-defense, a protection of the personality from intolerable violation. This amounts to saying that people have a *right* to go mad, since they have, in effect, been victimized and then mystified into abnormal forms of conduct.

Laing's more professional interest is in the intricate way families do such violence to their more vulnerable members. (Here he follows Reich, who also believed that the family is the central repressive institution, "a factory of reactionary ideology and structure." But in his more philosophical (and apocalyptic) moments, Laing has broadened the analysis to include the entire culture whose agency the family is. The implication of his work is that neurosis (especially schizophrenia) can be seen as a sign of rebellious vitality on the part of the victim; its strategies of compensation and evasion may even be ingenious. Thus, madness may be the sort of "disease" that pregnancy is: the beginning of a rebirth of the personality—provided there is a sympathetic environment which can act as midwife to the suffering soul. Therapy may then mean letting the illness run its course, until the person emerges from the other end of the tunnel, stronger and richer for the adventure.

From this viewpoint, the mad, like Dostoevsky's Prince Mishkin, play Christ to a crucifying world. They are those still human enough to suffer for the world's sins. Perhaps (to take the thought a step further) they are even objects of admiration, the martyred conscience of the age. Or so the more extreme interpreters of Laing's work—such as David Cooper—would have it. Certainly, the insane are not, in Laing's view, to be patronized and manipulated by arrogant analysts. Along with Thomas Szasz, Laing has made the most heroic effort to "demystify" and deprofessionalize psychiatry, to make the treatment of the insane a matter of patience and compassion: a sort of respectful attendance at the mad one's "experiential drama."

What flows from Laing's work is a forceful indictment of contemporary society for its murdering, if well-rationalized lunacies. This is what radical therapy has taken from him, and it is something the higher sanity must preserve. But for all the passion of his indictment, Laing, like the radical therapists, seems reluctant to press his case beyond the level of corrupted social relations. With the result that he almost appears at times to be laying the world's genocidal violence, exploitation, and ruthless egomania wholly at the feet of mothers and fathers who are obviously themselves as much victims as those they victimize. When it comes to the deeper sources of alienation, Laing has only very obliquely touched on the eroded condition of our transcendent energies. Very likely this is because the postwar existentialist philosophy on which he draws so heavily simply cannot support an adequate critique of our cultural condition.

Somewhere at the bottom of Laing's work is Jean-Paul Sartre's Stoicism, along with some conventional left-wing assumptions (which I would accept) about the general criminality of capitalist society. But Sartre's fanatically secularized humanism is itself the symptom of a still more lethal disease that grips all the modern world. And this malaise of the spirit will never yield its secret to the most involuted phenomenology of interpersonal communications (Laing's specialty)—unless, like Buber, we can see our estrangement from one another as a reflection of our estrangement from the sacred.

In some more recent and adventurous writing, Laing has come at this possibility cautiously, suggesting that the "breakdown" of neurosis (or of the drug trip) may participate in a transcendental "breakthrough." [4] It is an idea that finds a ready audience among those for whom blowing the mind by any means available is the shortest route to enlightenment. It is, once again, a sign of our advanced spiritual alienation that so many are willing to identify the paranor-

4. See his essay "Transcendental Experience" in Theodore Roszak, ed., *Sources* (New York: Harper Colophon Books, 1972), pp. 420–433.

mal and the abnormal with the transcendent, and so to confuse the real thing with its caricatures and counterfeits. Once make that mistake and one will next be in the market for electronic satoris, pushbutton epiphanies. And not much further on lies an indiscriminate infatuation with every psychotic and demonic outburst that happens along—one of the worst risks of the transition we are in.

To respect the mad as our fellow humans is one thing; to adulate them is quite another—and perhaps no better a course than to persecute them. Some saints behaved like neurotics; but it was not neurosis that made them saints. Our psychiatry still has much to learn from those primitive societies (I think especially here of the North American Indians or some Nigerian tribes) whose practice it is to bring all queerly behaving children to the shaman for examination; the child's eccentricity might, after all, betoken divine gifts. Ordinarily, after a few days of observation, the shaman can decide the case. Yes, he may say, the child is indeed blessed. Or no, the child is simply unbalanced—in which case, the boy or girl is sent home for tender, loving care.

Only a society much more deeply studied in the strange ways of the spirit than is ours can make such discriminations with confidence.

Human Potentials: The Sanity of Gratification

The Sanity of Gratification has so much to do with unrestricted experiment and improvisation that one might easily doubt that it is any form of psychiatry at all. Certainly its practitioners have violated professional conventions with a shamelessness that very nearly resembles contempt. Yet, in the form of the Human Potentials Movement (actually an expanding frontier of avant-garde therapies loosely linked together by a network of Growth Centers patterned after

California's Esalen), it has pioneered an astonishing number of techniques for exploring the full range of perception and consciousness; and from this has arisen a challenging new standard of psychic health and organic well-being which no school of psychiatry can any longer ignore without drastically limiting its understanding of human nature. Human Potentials deals in growth and adventure; its object is less to patch up old psychic scars than to liberate new possibilities of experience. Its work has little relevance to the treatment of the psychotically disabled; rather, it draws its clients from among those who are functionally sane but underdeveloped as personalities. Its practitioners are like doctors who are not content to set a broken leg but who want to go on to teach their patient how to dance. The goal is not simply to cure but to enhance.

The Sanity of Gratification might also be described as everything radical therapy, adjustive psychiatry, and Freudianism most despise. As against the radical therapists, it treats the individual in strict isolation from race and class as someone who has every right to purely personal growth and every possibility of achieving it. As against the psychic adjusters, it proclaims unlimited human freedom and the supreme virtue of originality. Working entirely from imaginative introspection, it seeks to awaken those spontaneous and even rebelliously creative qualities in people which are the bane of the adjustors and behaviorists. As against the Freudian tradition, it is adamantly antireductionist, insisting on the existence of higher centers of consciousness which can transcend the instinctual forces and liberate us from infantile hangups—and do so in startlingly short order.

Above all, and perhaps most obnoxiously from the viewpoint of all other psychiatric camps, the therapies of gratification do not hesitate to promise their clientele joy and personal fulfillment. Their style is shamelessly hedonistic and even playful. They will use any aspect of personal experience to achieve pleasure, humor, serenity, erotic delight. Art, song, eating, breathing, dance, free play, nature walks, athletics—all these have been brought into the repertory of the Growth Centers as means of self-knowledge and expression. "Bring anything that tickles you," requests one weekend workshop offered at Esalen, "a friend's uncle, an over-ripe banana, jokes, cartoons, personal incidents." And another: "Let us create a space to renew our sense of aliveness, moving freely in our own perfect and personal rhythm, chanting, dancing, eating, tasting, and soothing, caring and touching one another with massage."

If these light-hearted invitations make the Sanity of Gratification sound frivolous, one should understand that behind the bulk of its experimental work there stands an impressive genealogy: Wilhelm Reich's bioenergetics, Frederick Perls's Gestalt therapy, Abraham Maslow's humanistic psychology, Jung's wide-ranging studies in mystic and occult tradition. All this has become richly mixed with Oriental yogas and meditative disciplines, and with body therapies like those of Charlotte Selver and Ida Rolf. People are drawn to the Growth Centers for many reasons: some for fun and games, some for far more hazardous explorations of the self. All together, the curriculum amounts to the most significant departure in Western education since the Renaissance: a brave effort to break down the verbal-cerebral monopoly over the personality. "Affective education," as the work of the Growth Centers has been called, takes learning to a deeper level than it has ever achieved in our cultural mainstream. It reaches into the organic subsoil from which ideation grows: the nerves and muscle fibers, the bones and guts of us. As this style of instruction spreads its influence, it is likely that, within another generation, no one will regard anything as education that fails to reach so deep.

The vice of Human Potentials is that too often it desocializes its students, giving them the experience of everything . . . except the world's injustice and the more vulgarly material sufferings of their fellows. The encounter group and sensitivity workshop, the Gestalt session and Arica seminar, can easily distract from social criticism and moral obligation —by the very fact that their results depend upon work that is intensely personal. And at that point, unless we are very careful, we are close to unpardonable self-indulgence and an introspective dead end. It is notorious to what a limited high-income range the American Growth Centers cater. Why should their services at that social level not be classed among the more refined luxuries of the affluent society, tantalizing commodities available only to a moneyed few? How can those who dispense such expensive attentions be certain that their therapies are not, like all the extravagant amusements in our society, among the rewards that go to loyal high achievers?—a perfect example of what Herbert Marcuse has called "repressive de-sublimation."

The best of the Growth Centers are not unaware of their need to achieve social relevance, though without at the same time surrendering the many techniques of affective education which are uniquely theirs. A recent statement put out by Esalen finds a significant bond between politics and the expansion of consciousness. The "severe censorship of human experiencing," the statement runs, sounding strong echoes of Wilhelm Reich, "doubtless contributes to the dangerous dis-ease of our times. Cut off from the richness of the senses, body, and inner life, the nineteenth-century American or European turned to war and the conquest of the physical frontier. . . . The impoverishment of the inner life leads the twentieth century towards drugs, crime, and a compulsive grabbing for consumer goods and transient external experience."

While the statement lacks the sociological precision of a

good, radical indictment, it does show an effort by the Human Potentials movement to speak to the social evils of the age. And as a general diagnosis of our condition, there is much in it which the higher sanity can endorse. But the problem remains: how do we make *sure* that the expansion of consciousness Esalen and its counterparts seek is sufficiently sustained in enough people to cure the "dangerous dis-ease" of the times? Certainly an occasional weekend at the local Growth Center will not do the job—especially when one considers how varied the quality of the Centers can be and the weight of their offerings. The possibilities of trivialization—on the part of both the instructors and the students—are always great. The fancy prices involved are apt to be a corrupting influence; they must surely work to restrict the student body. And once the seminars and workshops break up, where are most of their participants but back in the world of "compulsive grabbing" again, there to stay until they can buy their way into another oasis of exotic experience?

Yet, if anybody is doing the job of affective education on a major scale, it is the gurus and therapists of the Human Potentials Movement. Who can say how far their influence might not reach and along what strange avenues? Certainly it can do nobody's radical politics any disservice to awaken in people, at whatever level they are ready to be awakened, a desire for uncensored experience—even if, for some, it is only a desire for authentic enjoyment in a world whose joys are shabby commercial compulsions.

Moreover, where the Growth Centers begin to exercise the visionary powers, they invite us to cross the boundary that divides self-gratification from self-transcendence. In increasing numbers, therapists who began their work at the Centers in encountering, Gestalt, or psychosynthesis have migrated across that line, turning for more and more of their insight and technique to traditional religious disciplines. As that happens, we enter a new and at the same time a very old phase of psychotherapy. We are on the borders of the higher sanity.[5]

The Planetary Dialogue

I have suggested that the higher sanity will take as its model of health the example of saints and shamans, prophets and seers—those who follow most faithfully in the path of the Few—and that it will borrow heavily on the teachings and symbols of the visionary quest. But there is no graceful way in which this borrowing can be done *directly* by modern society; what comes to us out of exotic tradition by simple appropriation or imitation is more apt to be a fossil or a museum piece than a vital commitment that brings our best, creative powers to the surface. Archaism and the quick counterfeiting of alien lifestyles are bound to be constant vices of our transition to a planetary culture. And, ironically, the more slavish the imitations, the more we are required to suppress our own experience as the world's first planetary generation, to play dumb about what we have uniquely learned along our way. At the very least, we know what only a cosmopolitan experience of the world can teach: that every tradition is only *one* tradition among many, that none can claim an exclusive validity. That fact alone must deeply color our understanding of the options before us, especially of those that demand a monopoly of conviction which no modern person can honestly offer. We must learn to be loyal to our choices without being parochial in our study or appreciation.

This is the major reservation one must have about Aquarian developments like Krishna Consciousness or Scientology or the Jesus Freaks. Their exclusiveness betrays the great responsibility of our time, which is to work toward planetary

5. *The Esalen Catalog* gives an excellent overview of new and continuing work in Human Potentials. It can be ordered from Esalen Institute, 1793 Union St., San Francisco, California, 94123.

synthesis by making ourselves the students of all human culture. That radical openness is at the heart of the higher sanity, a risky challenge not only for those who approach it from the parochialism of Western science, but for those who approach from spiritual commitments which demand a total, mind-crucifying allegiance. In the case of Krishna Consciousness, for example, we have a determined effort to import and imitate in exact detail an exotic religion which cannot help but become a bizarre curiosity in the streets of the modern world. There is no intellectual outreach, no desire to integrate and synthesize, no search for common ground. There is only radical and self-righteous disjuncture, plus a stubborn determination to convert the infidel. In the lives of young Krishnaites, the entire Western experience (or as much of it as they have acquired since childhood) must simply be switched off and obliterated by unquestioning obedience and rigid imitation of an alien model—to the extent that children born to members of the movement are now being raised at special, rigorously insulated schools which provide a hothouse environment of Krishna Consciousness. It is indoctrination pure and simple—the total subjugation of the young to the parental generation. As with the Jesus Freak sects, exclusiveness is the heart of the Krishna Consciousness Society. Which means that egoism is the heart; the worst sort of egoism—the desperate need to be uniquely and wholly right. No profession of humility can mitigate the sheer arrogance of such a claim to spiritual truth, and nothing but willful blindness will ever maintain it.

Of course, there is a place in our culture and in the higher sanity for those who convert to the lifestyle of exotic religions. Provided they use what they learn for teaching rather than preaching, their choice can be seen as a kind of existential scholarship. They can become major conduits for the "inside story" of alien spiritual traditions; they can offer us a perspective of total commitment which academic study may never achieve. Thus, though born a German and raised in the West, the Lama Anagarika Govinda has emerged as one of our best and most widely respected sources on Tibetan Buddhism, precisely because he can speak to us of his experience from a European viewpoint.

What the higher sanity requires of the philosophies and therapies that serve it is open, planetary dialogue between modern experience and sacred tradition. To either side there are vices that vitiate the dialogue. On the one side, there is the effort to affect an exact, pedantic replica of the traditional; to go this way is to deny the significance of our present planetary moment. On the other, there is the effort to force tradition into the secular categories of modern thought, either psychological, or sociological, or anthropological; to go this way is to deny the autonomous value of traditional knowledge and, indeed, the reality of the sacred. In Blake's words, it is an effort "to vegetate the divine vision." Somewhere between these vices lies the home ground of the higher sanity: a space large enough to contain the scholarship that brings us an accurate knowledge of exotic culture and ancient tradition, the nonreductionist scientific research that makes itself the respectful student of these sources, the art and philosophy that borrow insight from sacred teachings, and, at last, the therapy that seeks to recapture in experience the underlying vision on which those teachings are reared.

The balance is delicate, but it has often enough been achieved with grace and imagination. Over the past two centuries, the modern Western effort to assimilate the traditional and the exotic has grown into a tradition in its own right—dating from late Romantic fascinations with primitive and Oriental cultures. Goethe's brilliant adaptations of the Sufi poet Hafiz, Blake's amalgamation of biblical prophecy and the Hidden Wisdom, Wordsworth's spontaneous paganism, Baudelaire's use of alchemical and magical influences—all these are pioneering efforts to build bridges between the increasingly secularized West and the religious culture of the world, while at all times treating both partners to the dialogue as respectful equals, even where the one stands as pupil to the other.

All along the way, there have been false steps and foolish mistakes in this search for synthesis. Usually they have been born of the desire to achieve too much too quickly or to systematize too rigidly. James Frazer's armchair anthropology in *The Golden Bough* and Arnold Toynbee's study of world history are flawed by such vices, though they are heroic and inspiring labors nonetheless. In other cases, the results have been sloppy scholarship (as in the cases of Eliphas Levi and Madame Blavatsky), graceless congestion (as in Joyce's brilliant but overwhelmingly obscure *Finnegans Wake*), or plain, silly caricatures—like the almost burlesque occultism of improvised ritual in the Order of the Golden Dawn, or the many superficial forms of Orientalism that are no better than chinoiserie. Even so, it was the Order of the Golden Dawn that yielded the scholarship of A. E. Waite and G. R. S. Mead, and the later symbolic poetry of Yeats, one of our society's most successful assimilations of occult teachings.

In our own day, the dialogue has broadened and strengthened. We enjoy the fruits of first-rate scholarship on every aspect of exotic culture and sacred tradition, as well as increasingly direct access to the adepts of many Oriental religions, some of whom, like the Tibetan masters who have lately migrated to America, are making a remarkable attempt to graft their knowledge on to Western philosophy and psychiatry. As a result, we have had a number of noble efforts at intelligent integration. Aldous Huxley viewing all the modern world through a sensibility molded by Vedanta, Thomas Merton reaching out to Buddhist monasticism and the Sufis, Joseph Campbell's pioneering work in comparative mythologies, the metaphysical exploration of sacred tradition by the many contributors to the *Journal of Comparative Religion* (especially Titus Burckhardt, Seyyed Hossein Nasr, and Frithjof Schuon), the wise and harmonious eclecticism of Ezra Pound's and Kenneth Rexroth's poetry, Paul Goodman's blending of Taoist naturalism and Gestalt therapy, Gary Snyder's poetic assimilation of Zen and American Indian influences, the freewheeling hip-Hinduism of Allen Ginsberg and Ram Dass, Alan Watts's clever syntheses of Eastern religion and Western psychology, Dane Rudhyar's sophisticated adaptations of astrological lore, Jacob Needleman's efforts to open mainstream philosophy to the esoteric traditions, Jerome Rothenberg's rhapsodic translations of primitive poetry, Gershom Scholem rescuing Kabbalism from obscurity, Gay Luce's skillful integration of Tibetan mysticism and Western body therapies, and, of course, the many-sided effort of Carl Jung to find a generous place in modern psychiatry for the religious and occult heritage of the world . . . all this and much more than I can even list here has been part of the adventure of planetary dialogue. Nor has the enterprise been limited to intellectual and literary efforts. The painters and sculptors of the twentieth century have doubtless done more than any anthropologist to teach us the truth of the primitive worldview by borrowing heavily upon its style and sensibility; and no psychologist has done more to read the deep and enduring message of the myths than Martha Graham in her dance dramas.

How shall we characterize the special contribution such artists and thinkers make to the higher sanity? They share several qualities of authentic dialogue. While there is a studious care to their work, their stance is that of *engaged* men and women, seeking a wisdom they can themselves live by; they are after live options, not moribund research. Further, they address the alien and exotic openly, without the intention of dominating or denigrating. Rather, they are gifted with a lively eye for the humanly universal in all they encounter; and, finding it, they announce it in an authentically modern idiom which is free of jargon or scientistic reductionism.

Above all, they mean to take at its full philosophical value the experience of transcendence that lies at the core of what they study; and they do so because they know that here is the experience our dying industrial culture cries out for. This is the "relevance" they bring to their art and thought: their shrewd perception of the role that the visionary energies must play in both the spiritual and the political life of our time.

We will, in the final chapter, review the centers of cultural consensus which are drawing together to make up the ideal of a higher sanity. Here, I wish only to emphasize the sort of art and thought that contribute to an authentic planetary dialogue, because it is to such examples of intelligent and engaged synthesis, more than to any body of professional literature, that therapists of the future will look in their task of creating modernly viable forms of spiritual education. We may expect to see the psychotherapy of the coming generation take on more and more the role, if not the actual style, of the old mystery cults to which troubled souls turned not for adjustment or gratification but for spiritual renewal. There will be a generous place in the new therapies for the values of solitude and meditation, for rituals of rebirth and the art of dying, for rites of passage and experience of ecstatic communion. More than anything else in our culture, therapy has the chance to make these great universal themes catch fire in the lives of people—but only if it allows itself to be the respectful student of traditions that have read the human heart more deeply than any science.

9

Ethics and Ecstasy: Reflections on an Aphorism by Pathanjali

Reading through Pathanjali's *Aphorisms of Yoga,* I come upon the following teaching:

Yama, Niyama, Asana, Pranayama, Pratyahara, Dharana, Dhyana, Samadhi are the eight steps of yoga.

I note that the steps best known to Western practitioners of yoga—from asana and pranayama (the familiar athletic postures and breathing exercises) to samadhi (high trance)—are placed *after* yama. And what is yama? It is moral duty, defined by the aphorisms as absolute rejection of violence, theft, covetousness, lying, incontinence. Here is how Sri Purohit Swami explains Pathanjali's meaning at greater length.

Refusal of violence is love for all creatures, refusal of stealing is love for one's neighbors, refusal of covetousness is maintaining the dignity of oneself, truth-telling is maintaining the dignity of society. . . . All life is sacred, all life is one; no one has a right to question the sacredness of another, no one has a right to commit violence against another. The yogi who wants to find the unity of life, should not break that unity. . . . Men differ in temperament, character, environment, but they all stand on the one rock of Self, and when man commits violence on man, he commits it on himself. . . .[1]

Pathanjali's aphorisms are brief to the point of terseness. Yet every word counts; every instruction is a truth and a discipline distilling the wisdom of generations. And Pathanjali tells us that yama comes first, before we learn to perform the postures of the fish or the plow, before we learn to inhale on the left and exhale on the right. Before we mount to these physical virtuosities, down at the very bottom of the ascent, we encounter the first step: yama.

And I wonder, who can ever climb this ladder? Who will ever be able to surmount that first step? For within the terms of yama, we could write all the ethical philosophy and moral theology of the Western world. It embraces everything that sacred scripture and political ideology have ever demanded of us. Charity, sacrifice, the love of neighbor . . . they are all there. Liberty, equality, fraternity . . . they are also there. If we make the most of yama, we will easily find all the ethical virtues within its province. And must we not make the most of any great master's words?

There have always been those in the West who have condemned the Eastern religions—and especially yoga—for their lack of social conscience. Mention yoga, and for most Westerners the answering stereotype is an image of entranced ascetics blissfully tying their limbs in knots or contemplating their navels in the midst of famine, filth, and wretchedness, pursuing ecstasy among the heaped-up human carrion. Of course there are such world-despising yogis. Modern India is filled with them still, dropped-out, copped-out holy men, who stand before their society as emblems of heroic resignation. And there have been Westerners for whom that very image of impregnable tranquillity has been the whole attraction of yoga. They turn to the religions of the East to

1. Bhagwan Shree Patanjali, *Aphorisms of Yoga* (London: Faber & Faber, 1938), pp. 51–52.

learn the gift of tolerating the intolerable. Cosmic escapism, ethereal surrender. Still today the Transcendental Meditation of Maharishi Mahesh Yogi is merchandised to the millions as a way of adjusting to the modern world's abrasions. Don't resist; let your secret mantra wrap you in a cushion of private beatitude. Learn to float above the cries of outrage and anguish.

But I return to Pathanjali, and Pathanjali puts yama first: absolute rejection of violence, greed, deception. *First.* Not later. Not afterward. *Now.* By that single, abrupt dictum, we are reminded that yoga, at its most comprehensive, includes karma yoga, the yoga of moral endeavor, of sacrificial service to one's fellows. The yoga of Gandhi. And of Tolstoy, Keir Hardie, Danilo Dolci, Martin Luther King, A. J. Muste, Dorothy Day, Cesar Chavez. All these are karma yogis, men and women whose search for God led them to the compassionate struggle for justice. Their way, Pathanjali tells us, is not simply *another* yoga alongside hatha yoga, bhakti yoga, jnana yoga, which we may choose or not, as if the yogas were a smorgasbord spread before us from which we select whatever temperament or convenience dictates. If the root meaning of yoga is "union" (as in "yoke"), karma yoga teaches us that our union must be as much with the human as with the divine, and perhaps with the divine *through* the human. So Sri Purohit believes. "People forget," he tells us, "that Yama and Niyama [Niyama is a second set of disciplines which includes purity, cheerfulness, reverence] form the foundation, and unless it is firmly laid, they should not practice postures and breathing exercises."

I am reminded of the story of Shankara and the Untouchable. After having mastered the whole science of yoga—as he thought—and having attained nirvikalpa samadhi, the great philosopher discovered he had still one last lesson to learn, which should, perhaps, have been his first. One day, letting caste arrogance get the better of him, Shankara abruptly ordered an Untouchable out of his path. The man replied, "If there is only one God, how can there be any distinctions of caste between men?" Shankara, overwhelmed by both shame and sudden insight, fell at the man's feet to beg forgiveness and give thanks, saying "He who has learned to see the one Existence everywhere, he is my master, whether Brahmin or Untouchable." Thereafter, in his teaching, Shankara always dealt with caste privilege as among the most poisonous of illusions.

Before that moment, had Shankara really mastered yoga? Had he truly experienced the great unity? Could anyone but the Untouchable, in brotherly confrontation, have completed his education?

But if yama is indeed the foundation of yoga, how long does it take to build that solid ground beneath us? A labor of years, decades, a lifetime. For consider: how can any of us claim to be purified of violence, greed, and deception until we have driven these evils from the social order we reside in, live from, support by our votes, our purchases, our taxes, and our daily, silent acquiescence? The economy in which we get and spend is from first to last grounded in violence, greed, and deception. It extorts its wealth from the powerless and intimidates its stability out of the dependent. It thrives upon the limitless avarice of its stockholders and consumers. It sells all it makes by the grossest lying. Neither gentleness nor generosity nor truth is tolerable to the corporate leadership of that economy. And the tolerance of that leadership is not simply a disapproving scowl; it is an unrelenting, institutionalized pressure that works systematically to corrupt moral virtue. The military-industrial complex (the basis of the American economic system), the advertising industry (the basis of our high-consumption lifestyle) have ground violence, greed, and deception into the grain of our daily life, forcing us to accept the balance of terror, the engineered consensus, the lies of admass, as necessary and normal. We are the customers and employees of business elites for whom it is very nearly a law of nature that the company must never perform a generous act which is not a public relations gimmick or a tax dodge.

Hardly enough, then, to frown occasionally upon the ingrained and ubiquitous evils of our society and say we have, by that feeble gesture, "rejected" violence, greed, and deception. If that is what we would make of a great spiritual master's teaching—the least and easiest we can—then we may as well call a sloppy somersault a true shoulder stand; may as well lie flat on the floor and call it a perfect cobra posture.

There is no way around the challenge. The true yogi's first step is a moral step; higher consciousness is born out of conscience. "Consciousness"/"conscience": the very words are related, reminding us that we cannot expect to expand spiritual awareness unless we also expand our moral awareness of right and wrong, good and evil. Later perhaps there will be ecstatic harmonies beyond the description of words in which the good and the evil of the world will be revealed as, mysteriously, the two hands of God. But only the soul that has honestly cast out violence, greed, and deception may begin the ascent to that lofty vision.

Surely too many Western practitioners of yoga are playing trivial games with the psychic and physiological spin-off of the divine science. They learn to clear their sinuses, to mitigate their migraine, to flirt with the joys of the kundalini. Perhaps, besides achieving an enviable muscle tone, they even happen upon occasional intimations of samadhi. But all these achievements become barbarous trifles if we forget that yoga, like all spiritual culture, is a life discipline and a moral wisdom. While Pathanjali does not scold and bully us at length like the preacher in the pulpit, nevertheless the severe simplicity of his one aphorism makes his meaning clear to all but the willfully ignorant. On the way to samadhi, there are eight steps, and the first of these is moral conduct.

And again I wonder: if we take the instruction at its full value, who among us, in a lifetime of striving, will ever mount and surpass that first step?

But if we interpret yama in this way, must we then indict all the contemplatives and reclusive sages of history for criminal negligence?

I think there would be no justice in that. To do so would be to read the peculiar conditions of twentieth-century life into their historical moment, and so to lose sight of the ethical authenticity that once belonged to those who abandoned the world to seek God in lonely places. Because the urban-industrial civilization we live in is a world-girdling economic and political network, we cannot imagine escaping from it in any honest way. Its corrosive ethos pervades the planet. So we easily forget that the city and its imperious corruptions were once—until no more than a few centuries ago—an exceptional and strictly provisional way of life. Whole empires of the past might embrace no more than a handful of urban dwellers beyond whose walls rural villages and untouched wilderness dominated the landscape in all directions with their placid and nonexploitive routines. One *could* escape, break free, and live clean with one's guru on the locust and wild honey, achieve self-sufficiency in solitude and invite others to do likewise. It was not unrealistic to imagine whole cities being abandoned (as they frequently were in time of plague or famine) by inhabitants returning to the countryside or the forests to find freedom, safety, purity. The early medieval monasteries and convents induced nearly one-third of the Western European population to reject the world in favor of an alternative, self-supporting society of spiritual communes based on nonviolence, honest labor, service, and contemplation. It was once possible and even practical to imagine building a reasonable facsimile of the City of God somewhere just over the horizon or across the sea. There were still new lands and new worlds that offered the chance for a new beginning.

Perhaps this vision of a world where escape and innocence were possible still dominates the minds of the yogis and swamis of the East. Perhaps the influence of tradition clouds their understanding of how global the forces of violence, avarice, and exploitation have become. But we in the West, who generate so many of those forces, ought to know that the age of reclusive mysticism and personal salvation is past. Enlightenment has become a social responsibility; of necessity, we must strive as best we can to become prophets and bodhisattvas unto our society at large, prepared to see political action as a required sadhana. For where does one run today to deliver one's soul from perdition? How does one stop being part of the world's evils in a world that is hell-bent on becoming one planetary industrial cosmopolis? The wilderness is filled with missile sites and pipelines; the faraway mountains are being pulverized into fuel for the engines of Juggernaut; the deserts are filling up with oil derricks, test ranges, uranium mines. Can there be any hermit saints left on the globe who have not been staked out as tourist attractions? If so, they will surely soon go the way of the tiger and the whale: squeezed out, overrun. And where then will we find sacred ground and sanctuary?

If the purpose of yama is to define a fertile space where gentleness and truth can blossom and sustain the religious quest, this can no longer be a space we go elsewhere to find, but a space we must clear for ourselves, like a homesite in a thicket. It must be ringed by a line we ourselves draw and defend. We must commit ourselves to making yama a reality where we are here and now, making that struggle part of our growth.

Apparently, even the once adamantly antipolitical Society for Krishna Consciousness has come to recognize the unique conditions in which mystic religion finds itself and has decided to speak to the social wrongs and needs of the day. The Hare Krishna youth have lately launched an "In God We Trust" party to contest local elections. Not with the expectation of winning, but, in the tradition of protest parties, to educate the electorate to new values. The main plank in their platform: to withdraw 60 percent of the national wealth from violent, commercial, and materialistic uses and to dedicate it to the praise and worship of Krishna, primarily by building temples where, in the spirit of the Society, all people may come to eat, live, work, and enjoy a meditative stillness. A zany idea at first glance, especially given its Indianized veneer. But think of it as massive public works, combined with a free food and shelter program for the needy (plus sanctified peace and quiet for everyone), and it looks like as humane a use to suggest for our vastly wasted affluence as anyone has come up with.

As the children of a lawgiver God, we in the West cannot help but make moral duty the highest expression of religion. Even when religion fades from our lives, a special urgency, like that of the soul struggling for its eternal salvation, continues to cling to ethical conduct; the urgency may then even increase, for all that remains of spiritual aspiration in the life of the atheist concentrates itself in moral action, which then stands forth as the final evidence of one's authenticity. In this single-minded ethical emphasis lies the distinction and the glory of the Judaic-Christian-Islamic religious complex. When, therefore, we turn to the Eastern religions or to the religions of primitive people, we are appalled not to find moral conduct at the heart of their worship. Instead, we find there the pursuit of ecstasy, either by personal realization or by way of god-intoxicating tribal rites. It is not that moral conduct has no place in these religions. Don't the bodhi-sattvas pledge themselves to surrender personal well-being for the benefit of all sentient beings? Their instruction in the six traditional Buddhist virtues teaches them to honor compassionate sacrifice (*sila*) before reaching meditation (*dyana*) and wisdom (*prajna*). So too, as we have seen, Pathanjali, read with depth and sensitivity, makes moral purity the first step of yoga. But the terrible ambiguity of that "first step"! For what is "first" is preparation, not fulfill-ment. If we say morality is the "beginning," we cannot help saying that it is *only* the beginning. Beyond yama lie the discipines of ecstasy, the long, hard road to samadhi. And what an affront this is to the Western religious con-sciousness: that what is for us the summit of the spiritual life is, within other traditions, only a foothill.

The standard Western rejection of mysticism stems from this sense of outrage at the seeming casualness with which mystics confront the evils of the world, letting their attention wander to other purposes. With what righteous indifference they seem to hide from human suffering in the depths of the godhead. Now, again, in the adventures of the Aquarian frontier, many humanistic critics see that same moral failure spreading among us. They see conscience being subordi-nated to tricks of consciousness. They see the course of with-drawal, solitude, introspection, tranquillity becoming a wholly self-centered, psychic athleticism. Let me quote from a letter I have received from a prominent, politically in-volved intellectual leader. Commenting on the importance I have attributed to mystical religion in contemporary society, he complains that such "subjective thinking"

is like the dream of the counter-culturalists that, if only people went into communes and ignored politics, capitalism and society would vanish away. I know that politics, including radical poli-tics, is slow and painful and even dirty; but in my opinion it is the only way to fight—objectively. . . . It is easy to mislead the young with the illusion that immediate, subjective thinking is the key to reaching honest and moral solutions. What about the idealistic Nazi youth and Hell's Angels? You anti-rationalists will bear a grave responsibility for paving the way for irrational ideologies.

The easy equation my critic sees between visionary experience and "subjective thinking," and, in turn, between subjectivity and the Hell's Angels, betrays a sad if commonplace confusion; but there is nonetheless a stubborn moral concern to his accusation that I can only admire—and endorse. Certainly I would not deny that there are many who dally with mystical diversions out of weariness, boredom, or pure self-indulgence, looking for nothing more than the chance to be blissed out. With respect to such escape artists, I would not want to see an ounce of the humanists' moral pressure relaxed. For certainly the current generation of swamis and gurus, and their disciples, must come to see that the traditional postures of social withdrawal practiced by mystics of the past are no longer available for use—not with any degree of ethical dignity.

How tempting it becomes at this point to demand that the mystical vocation be politicized out of existence, that the ethical demands of yama be used to eclipse all that lies beyond. Certainly from the viewpoint of the conscience-stricken political activist, the disciplines of spiritual growth that stand to the far side of yama and to which Pathanjali gives so much attention appear to be nothing better than selfish distractions from social responsibility. Where introspection begins, society leaves off.

But in truth, this would do as much violence to Pathanjali's moral insight as the would-be yogis do who take yama at its lowest and least demanding terms. Escapist mysticism fails to recognize the indispensable place of moral action in the quest for God; but single-minded social conscience fails to recognize the place of transcendent experience in the quest for ethical commitment. So it fails to address the great question that must be answered by any ethical teaching: *Where shall we find the will to do what we know to be morally right? Why choose the good?*

The traditional answer in our culture has been either bribery or bullying. Christian moral theology lays before us the commandment of a despot God and then threatens us with "the loss of heaven and the pains of hell" if we should disobey. The rest is endless catechizing and preaching. So too the secular humanists who inherit from the God of the decalogue. They can make no appeal to the afterlife, but (and I think here of figures like Bertrand Russell, John Dewey, Jean-Paul Sartre) they continue to preach morality with the same towering confidence that people can be shamed and scolded into virtue. For all their rejection of religious tradition, the secular humanists stand like the Lord God upon Mount Sinai, thundering "thou shalt" and "thou shalt not," expecting to change the hearts of people by the raw force of impassioned rhetoric alone. Of course they do not often succeed. And so they conclude that people are not *rational* enough to be good, that the popular mind is too much swayed by superstition and primitive emotional impulse. If only people would be more *logical*, more dispassionate . . . if only they would get their emotions under sensible control. Pressed far enough, this line of ethical thought might soon bring us to the position of B. F. Skinner and the behavioral therapists, who would systematically train our stiff-necked society to be good by manipulating rationally devised "schedules of reinforcement."

But there is another answer to our question, one that we have from Pathanjali and the mystic sages of every religious tradition. They remind us of a vocation we have in life which makes the fulfillment of yama not a hard and grudging sacrifice but an action as gracefully spontaneous and inevitable as the organic need to grow and ripen. What they teach us is the completion of our identity, the achievement of that self we were born to become. And once having glimpsed that destiny, we mount to a height of the spirit from whose vantage point the sordid ambitions that tyrannize benighted minds sink into despicable insignificance. From that perspective, the worst villains the world has known stand convicted, not of an evil that makes us tremble with horror, but of a folly whose sheer banality tears the heart with pity. For indeed, "What does it profit a man if he gain the whole world and lose his soul?"

For both the socially committed Christian and the conscience-stricken humanist, all that goes beyond moral duty— the disciplines of mystic insight, the uses of solitude and meditation—is wholly dispensable. What both fail to see is that visionary experience is to ethical action as the soil is to the seed. It is the life-giving ground and sustenance of our moral growth. It is all that makes conscience something more than forced obedience to a code of ethics which stands over us like a scowling taskmaster. For this reason—because the light of visionary experience illuminates the meaning of moral duty and makes it the natural action of our humanity —I think the order of Pathanjali's steps is not necessarily or wholly a temporal sequence; it is rather an hierarchical order of experience. So yama becomes the inner circle of yoga which is, in turn, contained by others that give it meaning and cogency. But there is an interplay, a coming and going between the circles. Yes, there must be moral purity at the foundation of our lives; but it is only as we approach samadhi—gradually, steadily—that we come to see why this must be so. It is by our increasing mastery of yoga's advanced steps that we recognize the indispensable role yama plays in our growth. And, in turn, as we gracefully integrate yama into our lives, we manifest and confirm our progress toward samadhi. For what honest knowledge can there be of the great unity on the part of those who cannot even achieve a compassionate human solidarity? So there is between the lower and advanced steps a continuous rhythm, like the rhythm of the physical breath upon which we meditate in achieving pranayama. An interval of withdrawal and solitude, an interval of action and involvement—and each phase moves the next closer to perfection.

Perhaps this subtle reciprocation between ecstasy and ethics, vision and action, can be made congruent with the relationship between faith and works in those Christian theologies where faith is understood to be an experience, not merely a doctrinal assertion. An experience of what? Of the kingdom of God, which is eternally at hand and forever within us. What is yoga, after all, but a disciplined effort to find the kingdom and enter? The misfortune of Western culture has been its bad habit of substituting creeds, beliefs, and dogmas for direct experience of the kingdom; we try to capture in a trap of stilted language what can be grasped only by visionary awareness. Failing in that effort, we are left with nothing but words to serve as objects of belief, and nothing but threats of fire and brimstone to support moral conduct.

To be sure, there has been the Christianity of the monastic traditions which made meditative communion with God the hub of worship. So too the Christian alchemists, who conceived of the philosopher's stone as the vision of Christ returning in glory: they also possessed a well-developed mystic discipline. But Western history has not dealt kindly with these versions of Christianity; it has crushed them out or driven them to the cultural margin. The only experiential form of Christian worship that has grown and flourished over the centuries is that of the pentecostal congregations. But these have identified the experience of the Holy Spirit with any sudden and cataclysmic inpouring or raw feeling, any explosion of emotion whose source might be as much the subconscious as the superconscious. Historically, pentecostal religion has run either to antinomian excess or to puritanical repression, an ethical imbalance which suggests the confused quality of its experience. Though the Kingdom of God is within us, it is not *all* that is within us. There are also the pent-up angers and starved passions of a lifetime waiting to run riot through the mind. The weakness of the pentecostal churches has been their lack of spiritual discipline or tested ritual that can discriminate among the contents of the unconscious, telling the high from the low, ecstatic liberation from mere emotional abandon.

But that discipline is exactly what Pathanjali's yoga and many Eastern religions have created out of the collective experience of generations, with all the same painstaking care that we have devoted to science and technology. One might justifiably call visionary religion, and especially yoga, a science of the transcendent personality. And it is in such a science that we find the experiential base for a morality of nonviolence, truth, and justice.

We are, so I have already suggested, meaning-seeking creatures. We live to answer Tolstoy's great question, "Is there any meaning in my life that the inevitable death awaiting me does not destroy?" Where that question goes unanswered—worse still, where it cannot even find the language to express its existence—our lives remain unfinished, losing energy to a secret struggle against a vague sense of failure, as the half-person we are strains to play the part of the whole person we would be. That struggle may continue for a lifetime, weighing our hearts with despair, darkening our years, making death seem a more and more terrifying prospect—an all-devouring nothingness that renders our existence utterly senseless.

I will not say that such an unfulfilled existence makes nobility or goodness impossible. Clearly, people living under the most annihilating conditions of existential abandonment have nevertheless risen to magnificent ethical and creative heights. We know of the Marxes, Russells, and Camuses; how many more are there among the anonymous millions who have found their way to moral actions of astonishing heroism? But against what a dead weight of meaninglessness must these gestures assert themselves. Surely that weight takes its toll of our finest efforts, inhibiting the grace or purity of the act. Or it does worse; it drives us, in desperation, into the service of the world's parochial loyalties. Think of the millions who have with amazing courage laid down their lives in time of war. But for what poor, mad causes have they sacrificed themselves, too often in conflicts that now look to us to be a foolish waste. The very purpose of transcendent experience is to raise the mind above such partisan hostilities, drawing from them only what is universal as an object of loyalty, and sanctioning no means of defense which is not ennobling.

I understand the fear of fanaticism which inspires the modern conscience to practice a merciless skepticism toward all forms of true believing. The humanitarian concern that undergirds skeptical criticism deserves the highest respect. People in the grip of great causes easily turn murderous. But what is this if not a measure of how desperate their appetite for meaning can become—and a warning to those who would deprive that appetite of wholesome fare? "If God existed," the anarchist Bakunin said, "we would have to abolish Him" . . . the better to get on with the revolution. But Blake, no less a radical and no less a humanist, knew better. "Man must and will have some religion," he darkly observed, even if it must be "the religion of Satan." Militant skeptics who would have people imitate their own irreligious condition of lifelong, heroic doubt really ask the impossible. Not because people are too "weak" to bear the strain, but because they are being asked to deny themselves the experience of an insistent and magnificent human potentiality. It is as if the color-blind were ordering us not to see colors, insisting to us that to see colors is irrational and superstitious. So, instead of cultivating their talent for color vision gracefully, people hid it like a guilty secret; and when they could no longer stand the absurdity of doing that, they threw themselves into the experience in a spirit of angry or mindless rebellion that looked upon "rationality" as its enemy.

After two centuries of treating every fact and need of the spirit as criminal obscurantism, it is time for the secular humanists of our society to reflect upon their own moral origins. Where do they think the values come from which they use to condemn the aggressive bigotry of religious tradition and totalitarian politics? What is it in us that they appeal to when they exhort us to compassion and brotherhood? When we reject the fanatical, the inhumane, the parochial, it can only be by virtue of a higher loyalty that invites us to rise above the divisive passions of race, nation, culture, creed. A *higher* loyalty . . . the very language we must use to describe that allegiance speaks of transcendence, of the need to surmount and see beyond.

But that higher loyalty is scarcely a residue of simple good will and common sense. Nor will it reliably command action by way of pure ethical exhortation. It must grow from the conviction that we possess, in truth, an identity which expands the self vastly, making it more than *your*self or *my*self trapped in this single, desperate moment of history, mired in the tiny tribal hostilities that would claim eternal importance and enlist our whole heart and soul. To the degree that any of us transcends such transitory rancor, it is thanks to the visionary energies of the mind which have liberated at least a few of our fellows from the chains of time, allowing them to see a greater purpose to our existence than we shall ever find in domination, privilege, or power. They have taught us a quality of life by whose light we see the inhumanity that people commit against one another, not as infuriatingly wicked, but as acts of heartbreaking ignorance: a waste and a trivialization of our human opportunity.

What is the moral role of visionary experience? To make the mind big enough to show us the pettiness of the evil we would do to one another.

We have good reason to fear the need for absolute meaning where that need strays from the visionary powers which can alone keep it sane and kindly. Only the universal deserves to be the custodian of absolute meaning: nothing else —nation, race, class, empire—is capacious enough. There is more to give ourselves to than these lesser loyalties embrace; we serve them most becomingly when we know that we must, at last, outgrow them, recognizing how provisional a part they play in our personal destiny and in the evolution of human identity. We can do nothing nobler in behalf of these clashing secular allegiances than make an honest peace among them. But even that peace assumes its highest value when we see it not as an end in itself but as a means, just as the tranquillity of a contemplative interval is a means, a resource we use for our inner quest. We quiet the tumult inside our heads so that we may learn what the silence has to teach us. So, too, we quiet the self-righteous fury and violence that tear the planet so that we may get on with a far greater project than the politics of civilized societies has yet embraced: to awaken the god who sleeps at the roots of our human being.

And the project is universal.

"The Buddhas that have been, are, and will be, are more

numerous than the grains of sand on the banks of the Ganges."

So let us hope.

A half-century ago, in the midst of the First World War, Freud raised the great question: What is the origin of the explosive discontent that continues to defeat the purposes of reason and to torment the course of human affairs? To raise the question is to ask, at the same time, what the essential energy of history is. What is it that drives us forward in this violent and unhappy adventure? What is it that has been withheld from us and goads us into the inhuman conduct which has been the story of our time? In effect, Freud had issued an invitation to explore the demonic in our political life, the unassuageable craving that makes us the mad, genocidal creatures we are.

Answering Freud's question requires an archaeological study of human nature, a "dig" that will take us down and down through the many layers of need and motivation that define our being, as far down as imagination and daring will take us. The uppermost layers are by now familiar and very nearly accepted beyond dispute: bread, justice, equality. These needs have long since been integrated into the dominant ideologies of the modern world. They are well spoken for by the forces of liberalism and social democracy. Freud, in his turn, added the sexual needs of the subconscious and the even deeper need to express aggression, which he wisely traced to the fear of death. The existentialist philosophers and psychologists have added the need for authenticity, for assuming full personal responsibility for one's ethical commitments and one's fate. The New Left of our own day has added the need of a humanly scaled community and of direct participation in political decisions.

Quite as important as the discovery of these many new needs has been the growing realization that they are *orga-*

nically and *hierarchically* related. They cannot be satisfied in a segmented and piecemeal way, any more than the human organism can be healed as a number of discrete parts divided up among specialists, the stomach here, the psyche there. The medicine of the body politic must deal with us as whole persons, or its best intentions will go awry. Yes, we must eat and be housed and clothed. But if these needs are met paternalistically and without respect for our individual dignity and style of life (as is the case with most American welfare programs), then the result is hostility and demoralization. Yes, we must participate in the politics of our community. But if that social responsibility crowds out our need for some quiet and private space in which to be uniquely ourselves (as is the case in most people's republics and totalitarian states), then the result will again be festering discontent.

The problem is: we—by which I mean the intellectual consensus of the age—discover each of these needs on a different historical horizon. They seem to unfold out of one another sequentially. And as each is discovered, it seems to its finders to be the ultimate political value. So it is enshrined in an ideology or becomes the shibboleth of a movement, and is elevated above question. Anyone who then challenges the sufficiency of the ideology or the movement can only appear to be a traitor to the cause for having doubted the obvious truth that is here being championed. And none is more vulnerable to that accusation than those who speak in behalf of the spiritual needs. They must challenge what all the ideologies and movements of the past have in common, and that is their commitment to the great secular consensus which holds that all human needs can be fulfilled within the world of time and matter. All the dominant ideologies belong to that consensus; that is why their critique of the crushing dynamism and colossalism of modern industrial society is radically flawed. They cannot see how they, out of their own best intentions, have contributed to those life-denying forces.

This is no easy accusation to make when one is dealing with ideologies and movements that were created to serve mankind and whose adherents have sacrificed bravely and often. But the issue must nevertheless be joined. The needs of the person hold together organically; violate them at one level, and the result will be corruption and distortion all along the line, with the result that the worst evils—those which stem from deep and grievous discontent—will re-emerge like a malignant growth whose root has not been found. This much we should by now have learned from the history of modern revolution. Yes, the revolution may redistribute the benefits of industrialism more justly. But the same tribal hostilities continue to flourish under its banner; the same diseased preoccupations continue to surface as if by iron necessity: the bomb, the assembly line, the mass market, the production schedule, the rape of the environment, the ministry of propaganda, the secret police, the power-political rat race. . . .

Can we any longer believe these evils are simply due to the machinations of a few wicked cliques at the top of the social order? That a change of regimes would be enough to cure the ills? True, a revolutionary transfer of power may move a society one layer deeper into the hierarchy of human needs, and to that extent, the revolution will be at least a temporary advance . . . until the next level of discontent announces itself and all the evils born of frustrated need begin to reappear.

All is bound to go wrong with revolutions that work within the secular consensus because, at last, the secular consensus is wrong. It does not go deep enough to touch what is most fundamental in human nature, and so it cannot understand our discontent or bring us fulfillment. Here is where the Aquarian frontier becomes an essential part of contemporary political life. Those who rise to this historical horizon have faced up to the demonic in human affairs; they have been willing to learn about its nature from traditions and sources long since dismissed by the tissue-thin psychology of the modern ideologies; they have plumbed the layers of human need and motivation that lie further down than bread, sex, justice, participation, and have learned anew a very old truth, the truth that Tolstoy asserted against Marx, the truth that Jung asserted against Freud: that we are religious beings down to our very core; that there is no wholeness, no sanity, for us until we make spiritual need even more fundamental than all the others.

Trotsky once spoke of the need for a "permanent revolution," by which he meant a revolution that would spread from nation to nation until it had girdled the globe with socialism. A great battle cry, but what if that revolution should end only by producing a planetary 1984 or a world steeped in the *angst* of an Ingmar Bergman film? There is another sense, however, in which "permanent revolution" is exactly what we need: not a revolution that merely moves geographically over the planet, but one that moves along the depth dimension of human nature, feeling its way deeper and deeper into our needs and potentialities, refusing ever to say it has once and for all defined this unfinished and evolving human animal.

Because the secular consensus and all the politics connected with it stop short of our spiritual needs, at most devising anemic secular substitutes for what the visionary energies can alone supply, they progressively enlarge the spiritual void in our lives. And that void is the prime political fact of our time. It is the secret of our discontent, the anguish that animates our psychopathic conduct. The strenuous and foolish things that people in our time seek to do with history

—to multiply thermonuclear overkill endlessly, to raise up economies of limitless growth, to build conglomerate empires that straddle the globe, to turn the planet into one, vast industrial artifact, to produce without limit, to consume beyond all sane need, to propagandize the world with one's ideology—all this is what people use to fill the emptiness inside them. So too the mindless mass movements to which they surrender themselves in desperation; these also are among the corrupted stuff they cram into the void.

But the void is too big to fill. It is an emptiness that can be occupied only by something greater than all history and material achievement can offer. We are in the position of someone with a teaspoon trying to fill the Grand Canyon from a child's sand box. The task is hopeless; and because we know it is hopeless, we grow ever more frantic in trying to carry it through. Because we can see the thing we most dread and wish to bury—our own despair—rising out of the void like a man-eating monster. *Despair* . . . that is the demon of discontent within us.

It is despair that drives us into the maniacal history-making which is the hallmark of our age. The leaders we cheer are those who do the most to distract us from that despair. They hold great conquests, great hates, great appetites, up before us and cry, "See here! See here! *This* is how we shall pass the time! *This* is what we shall keep busy with!" An arms race . . . a bigger GNP . . . a new war . . . a society of ever more gargantuan metropolises . . . a great national crusade. To escape the void within, we even launch astronauts into the void of outer space, undertaking a "space race" which is a criminal waste of brains and resources. But we pay the price, because the spectacle affords us another distraction.

No, not all the crusades and projects are to be condemned. Some achieve humane ends. But more and more they become desperate assertions of national egotism, collective power, aggressive arrogance. As if we wanted them to achieve more than they can. And that is indeed what we are after: a gratification of the spirit from that which is below the level of spirit. The major cultural activity of high industrial civilization is turning out to be the invention of secular religions, bad substitutes which appear to wear thin within two generations.

Yet, behind the fanatical bustle of the modern world, the simple truth stands firm: people are not power- and profit-seeking creatures. Not fundamentally. Power and profit are without signifiance for the whole and healthy person. They become goals only by default and to the degree that higher purpose withdraws from our lives. That is why, since time immemorial, people have treasured above all else the memory of their saints and sages—the gurus who have turned aside from money and domination as if these were the rubbish of life. By their very presence among us, they remind us of the spiritual vocation to which we must, sooner or later, return.

Yes, people have other heroes too, conquerors and kings. But their rank has always been far beneath that of the gurus. History and legend preserve innumerable tales of saints chastising and humbling kings, but few of kings besting holy men. It was because they recognized their inferiority to the saints that kings and conquerors once kept jesters at their courts: to remind them of the essential folly of their careers. Clown and guru are a single identity: the satiric and sublime side of the same higher vision of life. Both reveal the achievements of the mighty to be vanity and vexation of spirit. Both oppose to the drunkenness of power, the intoxication of ecstasy. For what, after all, is the laughter a good clown brings us but the giddiness that comes of suddenly seeing— as if from a cosmic viewpoint—the absurdity of what the mighty are up to? For that moment, we taste the sanity of divine madness and become, for as long as the joke lasts, Fools of God.

These days our leaders are as little tolerant of clowns as of gurus. They take themselves *very* seriously. And *we* take them very seriously. Pity and laughter are what the world's ruling elites deserve. But pity and laughter are not possible as long as we occupy the same level of consciousness with those who rule us, or can regard them only from an envious worm's-eye view. We may hate and revile them on their own ground; but pity and laughter require a different ground to stand upon—a place apart and *above*. And there is the great problem. Modern culture—the culture of scientific and secular humanism—stands on one dead level and monopolizes all the ground. Denying the reality of transcendence, it insists there is no place outside and above itself—no sane place.

Yet, suppose by a trick of the mind, enough of us should do the impossible and find such a place to stand: a higher sanity. Then we might imagine a politics of pity and laughter which bravely revealed the emptiness at the core of our compulsive history-making; a politics that joined ethics to ecstasy to teach of better purposes, better destinations. An insurrection of the clowns and gurus, in behalf of their strange, beautiful, and transcendent sanity—*that* would be a revolution to match the need of our time.

Have we the imagination to invent such a politics of the higher sanity? What would it look like? Obviously, it would not take the shape of party platforms or official policy. Rather it would be a sensibility at once pervasive and peripheral, a critical awareness always alert at the edges of our public life, persistently counseling priorities and limits. From that sensibility we would learn that our lives aspire to a greater size than power or plenty in themselves can ever bring us, that these are, at best, a soil we prepare for further growth. We would learn to want just enough of both, which is perhaps not more than free labor, egalitarian sharing, and a healthy environment will readily yield.

And we would learn that we can never do justice to one another as social animals until we have each done justice to the visionary powers within us.

10

The Centers of Consensus: Reconnaissance of the Next Reality

Material Simplicity, Visionary Abundance

There is a sixteenth-century alchemical woodcut which has lately appeared in several "new consciousness" books. It shows a young pilgrim at the edge of the world, poking his head through the shell of the cosmic egg to glimpse in amazement the operations of the *anima mundi* that lie beyond.

These days, with the shell of our known world falling into a thousand pieces, a great many of us have our heads into various strange and astonishing spaces. Some see nothing beyond the limits of the familiar universe but an all-devouring void; and so do I sometimes. Others catch gleams of beatific realities on the far side of the apocalypse; and so do I sometimes. It is not always easy to tell prophetic fevers from delirium. But affirmations amid the fear and trembling are rare enough to be searched for any residue of hope they may carry—if only the reminder that the edge of one world may be the frontier of another.

There are pilgrims along the Aquarian frontier who hint eagerly of a future filled with psychic prodigies. They conjecture a world where all the wealth and well-being which urban-industrial society has pursued by technical cunning and physical force will become ours by the operations of higher consciousness. They are not cranks or science fiction addicts, but reputable professionals—academics and scientists of lively imagination—who discern in the untapped energies of the mind the promise of an ethereal technology, a "psychotechnics" free of material dead weight, drawing on resources of will and pure intellection whose only economic law is that of infinitely augmenting returns.

To mention a few examples that have come my way only within the last six months. A prominent psychotherapist remarks to me over lunch that people may sleep and die only because they have been mistakenly "programmed" to believe they have to . . . and goes on to suggest how the programing might be therapeutically redone to avoid both necessities. A neurophysiologist tells me of her research in liberating latent mental controls over pain, infection, and aging. A psychologist shows me photos of himself being operated on by Philippine psychic surgeons whom he has seen penetrate his body with their bare hands to remove cartilege and tissue. I attend a lecture where another psychologist tells of his promising experimentation with out-of-the-body phenomena. I come upon a physicist writing in *Physics Today* about "imaginary energy" and the proven possibilities of telepathic communication and precognition. A Tantric scholar whose work I am reading announces that Tibetan adepts possess the power of *lung gom*, trance locomotion at the speed of thought. In another recent book I pick up, a historian reports his belief that we can, by altering consciousness, plug into the power points of the Earth's etheric field and by so doing move matter and control evolution. An engineer I meet at a party explains how we might influence the Earth's geomantic centers and telluric currents by mental manipulations—which is the "technology," so he thinks, that built Stonehenge and the pyramids.

Perhaps all this is simply the higher gullibility of our day. But I am impressed by the cool conviction and purposefulness of those who survey these psychotechnic vistas and so tend to relax my doubts in their presence. It is too unkind to exercise criticism upon those who are giving their imagination a holiday. Their brainstorms are also part of the Aquarian adventure.[1]

Skeptical considerations aside, I am of two minds about their fascinations. On the one hand, I am left as cold by psychic as by technological marvels, believing that neither contributes much to the soul-saving wisdom which is the prime need of the time. No one levitates his way out of the human condition, even if he can in fact levitate. There are no parapsychic shortcuts to enlightenment.

But, on the other hand, whether it is a realistic proposition or not, I can see in this spreading interest in psychotechnics one important (and highly characteristic) way in which our society begins to feel out the contours of a welcomely new cultural possibility. It is only a beginning, and hardly the course I would choose to make the exploration; but through these bizarre speculations we see, if only in hazy outline, the prospect of *a culture of mind*, whose main work and play would be the elaboration of our underdeveloped psychospiritual capacities, the pursuit of transcendent being. And that interests me intensely. For certainly the time is at hand for us to begin internalizing as experience what we have for the past two centuries been externalizing as physical power.

1. Lyle Watson's *Supernature* (New York: Doubleday, 1973) is a popular compendium of psychotechnic wonders, and typical of the spirit in which many switched-on scientists and academics approach these marvels.

Time to see in the proliferating material stuff of our culture—
the vast machinery, the great industrial systems, the world-
shattering weapons, the mountainous stores of merchandise—so many waste and alienated concretions of our
visionary energies. We live in a world where the transcendent impulse has been deflected, trapped, and congealed
into *things*, into millions of things. No way to recapture that
dissipated energy but by disenthralling ourselves from things
and returning to the visionary origins of culture, the source
and substance of our higher identity.

The psychotechnicians may be a first faltering baby-step
toward that new culture. They are coming to see that the
machine has been inside the ghost all along, and is, in fact,
an *idea* of the ghost. At the further reaches of modern science, even the far-out physics of the day comes round to their
support. Professor A. D. Allen, writing in *Foundations of
Physics* (December 1973), suggests that the fundamental
stuff of the universe may not be objects of any kind, but the
nonmaterial principles of natural behavior themselves, like
the thoughts of a mind at work—a concept which, he admits,
is "almost theological in character." And Arthur Koestler,
in his recent *Roots of Coincidence,* rehearses the new directions in theoretical physics that indicate a model of reality
patterned after the mind—a current of research which he
is convinced flows directly into parapsychology.

I have already conjectured that the original human culture—the culture of the Few—was one of material simplicity
and visionary abundance. A sacred culture in which myth,
magic, and mystery were the prime preoccupations, and
qualities of experience the principal wealth. Is it toward
such an etherealized condition of life that sensitive spirits
now reach out, as the the urban-industrial dominance
lumbers toward collapse? If we try to imagine that primordial culture, the picture that no doubt first comes to mind is

that of a backward, tribal society, surviving meagerly by the skin of its teeth. Of course, it may not have been that way at all; but, in any case, our return to the source need not be a reversion to paleolithic austerity. Rather, it might be a discriminate adaptation of primitive simplicities at a new historical horizon. The burden of primitivism has always been its vulnerability to natural hardship and grinding penury; its vice has been its isolation and tribal divisiveness—all traits that cannot help but to wear down the finest qualities of spirit. It is to the planetary scale and a dependable plenitude that human culture aspires in its search for unity and universality. And these, it seems, can only be ours after the long, troubled labor of global integration which Western society has thus far advanced.

Material simplicity, visionary abundance. Who can say what gifts and wonders such a cultural style might not bring? Perhaps within a re-sacralized culture that lives by that standard, there do indeed lie all the astonishing psychic capacities we have been taught to regard as fantastic. Secrets of healing and longevity, of star roving and dream voyage, of extrasensory communication, teleportation, psychokinesis . . . powers that may, once upon a time, have generated the folklore of sorcery. Perhaps . . . perhaps. How much of that folklore is memory, how much make-believe? How much of what we make believe is real might not be *made* real, if we could liberate the transcendent powers within us? We do not know the limits of the will, only its weakness. And its weakness begins where our vision darkens. "Impossibilities" may only be realities vibrating in a wavelength of the mind we—or most of us—have not yet learned to tune in. Perhaps nothing we have ever imagined is beyond our powers, only beyond our present self-knowledge.

In any case, we are in no position to pontificate on the limitations of the visionary energies until we cease trying to counterfeit their purposes in time and matter.

The Centers of Consensus

We live among rumors and anticipations of the miraculous. But I will close on a more modest note. I will not try to predict the kingdom and the power that lie on the far side of urban industrialism, but only to mark out some part of the path that may carry us across the great cultural divide ahead. The task of our generation is to be such pathfinders of an evolutionary transition. Even those who cannot envisage our situation in evolutionary terms must recognize the magnitude of the challenge we confront in liberating ourselves from the death grip of the urban-industrial dominance. Nothing less than a revolution of the sensibilities will serve our purpose, whatever social revolutions we may also have to undertake. The way forward is inevitably the way inward.

Can we characterize that liberating sensibility? I think we can at least discern, among the thousand experiments that fill the Aquarian frontier, a number of themes which act upon the adventurous thinking of the day like gravitational centers. Taken together, these centers of consensus begin to describe a new reality principle uniquely suited to the spiritual needs of a planetary age at the troubled close of its urban-industrial phase. Conceivably, any number of religious or philosophical systems could arise to fill those needs—just as, in the early Christian era, any number of savior cults stepped forward to offer the promise of solace, absolution from sin, and personal immortality that were the crying need of that time. I seek here only to sketch the skeletal shape of the emerging consensus, not the worldview that might flesh out its bones.

My inclination, however, is to conceive of the worldview as a "therapy," if only because the popular vocabulary of inwardness in our culture has become the private preserve of psychotherapy. Even as we come to learn the essential oneness of therapy and religion, most of us approach that truth from the therapeutic side of the equation. It is psycho-therapy which serves as the major conduit in our society for all forms of consciousness exploration. If, as I have suggested in these pages, we are witnessing the evolutionary opening of a new endopsychic category—that of the higher sanity—it is therapy that is working most ambitiously and eclectically to meet that potentiality in the experience of most people. In the process, of course, psychiatry is itself being transformed by all it touches and borrows into a syncretic, salvational discipline. What follows are, I think, the rudiments it will synthesize into the next reality.

1. *Potentiality.* More and more, it is Pico's chameleonlike image of human nature that dominates the advanced therapeutic thought of the day. We are coming to see our-selves as an unfolding and perhaps infinite potentiality: the greatest show on Earth, an endless parade of curiosities and surprises. So we see the concepts of "growth" and "evolu-tion" becoming ever more central to contemporary psychia-try. People who pass through therapeutic experience easily perceive their personal history as a succession of roles and experimental identities they may change in midstream as self-knowledge deepens. With Blake, Nietzsche, Sri Auro-bindo, and the occult evolutionists, we have come to assume a fundamental human energy that needs to expand and ex-plore the possibilities of being in all directions, but ultimately in the direction of spiritual enhancement. As a species, we are not a fixed point in the universe, but an upward tra-jectory.

Even sexuality, once so dominating a preoccupation in Western psychiatry, now begins to take its place as only one stage of the growing process, one provisional use of our vital energy which must undergo maturation and perhaps, at last, be assimilated to higher growth needs. We begin to understand that the erotic imagery which appears in the world's religions as the metaphor of divine union is a deep recognition of sexuality's occult meaning. Far from religion's being an "aim-inhibited" manifestation of sex, it is sex that functions as a sublimated, physical expression of transcendence, an early approximation in our lives of a higher goal—the experienced unity of metaphysical polarities.

Once we accept transcendent potentiality as the essence of human nature, it becomes clear why material abundance and physical power, no matter how vastly multiplied, still leave us ungratified, restless, genocidally violence-ridden, or perhaps worst of all, plain bored with our existence. It is because these are, at best, means to an end; and, ironically, an end which may require that we finally outgrow the appetite for affluence and power we have struggled so long to satisfy. We finally understand the discontent that so persistently warps the lives of people and turns them ugly, even when they would seem to have every material blessing they desire. It stems from the thwarting of our paramount growth need: the need to evolve beyond the restrictions of time, matter, and mortality.

Our future image of human being, then, will be a strange, tense blending of the optimistic and the tragic: a study in paradox. We are optimistic in that we assume, not a radically "fallen" human condition, but a whole and healthy nature at the core of us; not an original sin, but an original splendor which aspires to transcendence and succeeds often enough to sustain a godlike image of human being. We are tragic in that we see how easily, in our chameleonlike freedom, we misdirect that energy toward lesser goals, unworthy objects. The psychotherapy of the future will not find the secret of the soul's distress in the futile and tormenting clash of instinctual drives, but *in the tension between potentiality and actuality*. It will see that, as evolution's unfinished animal, our task is *to become what we are;* but our neurotic burden is that we do not, except for a gifted Few among us, know what we are. What is most significantly and pathologically unconscious in us is the knowledge of our potential godlikeness. And, like Freud's repressed libido, that buried knowledge nags at the corners of the mind, demanding the energy it has been denied—until we grow sick with the guilt of having lived below our authentic level.

Even so, because our outlook is therapeutic rather than theological or reductionistic, our pathological tendency can be taken as a sign of our basic health. A psychology of potentiality is a psychology of dramatic development, of conflict and resolution: an effortful unfolding of the self with many a surprising twist to the story. With Nietzsche, we see that the human being is, inevitably, the "sick animal." Inevitably, but not incurably. As all spiritual disciplines recognize, the will to transcendence may often concentrate itself in "the dark night of the soul." Disease means psychic tension, and psychic tension is the potential energy of the spirit.

2. *Upaya.* We are leaving behind all religions, philosophies, and worldviews that exhaust themselves in language —in doctrines, catechisms, abstract analysis, academic head trips. From the Eastern and mystical traditions we have learned that true therapy must be an *experienced* knowledge, not a report or a conjecture, not merely a well-phrased idea or enunciated belief. It must be a rhapsodic fire of the mind rather than verbal ashes on the lips. We are learning how to stop being compulsive *explainers,* forever seeking to wrap the world up in words and deft interpretations. Even the Christianity that has of late become prominent among young believers—that of the Jesus Freak sects—prefers emotional enthusiasm to the usual doctrinal obsessions. It is theologically shapeless and rigidly fundamentalist, but filled

with pentecostal exuberance, a faith that means to be *felt*, even if the feeling is pathetically lacking in discriminate discipline.

In the Aquarian consensus, it is experience that matters: direct, personal perception that vibrates through the organism, transforming perception and conduct. At first, these affective upheavals may be sought in the raw shock power of narcotics, orgiastic sexuality, rock music, or—as with the Jesus Freaks—in stereotypic camp-meeting abandon. There are clearly many in our society who live so much of their lives in their heads—amid plans, calculations, worries, regrets, secret manifestoes, taunting self-images—that they have lost the power to feel. Everything they encounter becomes a verbalized fiction; their lives are without vivid immediacy. So, at the outset, they resort to such blunt sensational instruments to crack the body-and-emotion armor in which they have locked themselves. And anybody who gives them the chance to break loose immediately looks like the veritable messiah.

But, in time, it becomes clear that unstructured turn-ons and amorphous mind-blowing only add to the chaos of one's existence. They punctuate life with random flashes of ecstasy that rapidly burn away, leaving the mind as dark as ever, if not increasingly distraught: a downer after every upper, a bummer after every high. At that point, one must learn that there are *structures and disciplines of experience* as well as of intellect, and that these can absorb a lifetime's learning. Even the orgiastic rites of tribal and pagan societies were designed rituals, whose goal was sacred communion by well-tested means.

In the Vedic and Tantric traditions, such means—proven techniques for awakening and controlling visionary energy —are called upaya, and we are taught that upaya is the inseparable companion of prajna, enlightenment. If doctrinal hair-splitting has been the special fixation of Christianity, the multiplication of upaya has been the special labor of the Eastern religions—to the point that their mysticism (unlike that of the West) has become scientifically methodical. This is why the Eastern religions have developed the psychological sophistication to cope with and clear away many superficial hangups—like sexual guilt—which easily become lifelong agonies for Christians, many of whom cannot tell the difference between visionary experience and neurotic seizure.

Now, in its own rough and ready way, Western psychotherapy has begun to adopt the Oriental upaya and to improvise its own: exercises, meditations, sadhanas of self-exploration whose purpose is to outflank the verbal defenses of the ego and to penetrate the vital centers of conviction. By way of such affective methods, we are discovering the deep, transformative levels of learning which ordinary instruction rarely touches. Even many nontherapeutic subjects in the mainstream of education reach out toward these affective techniques. It is a sign of the times that the "workshop" has begun to take its place alongside the lecture and seminar as a standard teaching format. Once reserved for the manual arts, the workshop, with its games and exercises, now appears in all fields where participative learning is the goal, almost as the academic extension of group therapy. And surely where our purpose is to awaken spiritual potentiality, upaya—experiential technique—is of the essence. The article of faith, the authority of scripture, the creed learned by rote . . . these will no longer satisfy as a foundation for religious commitment. If faith is "the evidence of things unseen," it must nevertheless become *evident in experience,* if it is to be something more than a conditioned linguistic reflex.

It is worth remarking that already a therapeutic consensus has grown up among the new eupsychian techniques, especially as they fall under the influence of the borrowed upaya.

They come to focus more and more on *attention* and *relaxation*—or perhaps we can combine the two into a seemingly paradoxical kind of consciousness called "relaxed attention." The effort is, first of all, to shift attention—out of the head and its verbal abstractions into the here-and-now immediacies of the body, the senses, the visionary imagination. Secondly, one relaxes the imposed rigidities of body and mind, so that things may resume their flow and plasticity. The essential discipline of all the meditations and therapies has to do with "easing up" and "calming down," as if our main problem were that we habitually fight some natural gravity of the personality, stiffening against it, resisting, arguing with it, casting our thoughts before and after: bodies full of tension, heads full of words. The implied teaching seems to be that we are constantly distracting ourselves from a basic spiritual normality, forever refusing to let enlightenment happen to us—no doubt for reasons we would insist are excellent, even commendable . . . if only these gurus and therapists would let us *explain!*

In the future, as both religion and therapy invest themselves more heavily in upaya, we are bound to see a heightened appreciation of myth, symbol, and ritual in our culture, not simply from an academic viewpoint (that is already well developed) but as indispensable means of unfolding the transcendent potentialities. For these are the "texts" on which upaya draws in its labor of affective education. We are also apt to see the role of the guru, the master of upaya, reappear as the most honored of educational offices after long absence from our culture. Already the scene is crowded with imported swamis and home-grown "psychic facilitators" who hasten to supply the experiential pedagogy that book learning and conventional instruction cannot offer. So great is the demand for their services that upaya stands in danger of becoming a mass-processing operation—as with Transcendental Meditation and the Maharaj Ji's Divine Light Mission.

Or, in the form of group therapy, we have Erhard Seminar Training which treats "groups" in marathon sessions of two hundred or more at a time. The scale of the need is far beyond anything the mystic and psychic disciplines of the past have ever had to cope with. How to preserve the essentially personalist guru-chela relationship in such a situation without either institutionalizing it in a new church or turning away needy thousands?

3. *Transpersonal Subjectivity.* It was Jung who most prominently introduced myth and religious symbolism into modern psychiatry in a nonreductionist way. But even he, as great as was his respect for sacred tradition, never successfully clarified the status of the universal religious images he identified as "archetypes." Were they real or "merely" psychological? Even if one traced them to a "collective unconscious," were they wholly within the structure of the mind, or did they, in some sense, reflect a reality that went beyond human awareness? Were they simply the stored-up imprints of historical experience or visionary emblems of a transcendent origin? The epistemological issues were agonizingly difficult; intellectual respectability required a clear choice between subjectivity and objectivity, between In Here and Out There. Jung resorted to reticence and ambiguity, doubtless because his personal convictions drew him toward a religious commitment he was long loathe to publicize. But his usual tendency in public discussion was to psychologize the archetypes into complete subjectivity—and then to plead for their importance as a basis for psychic health. Indeed, there could often be a chill and academic abstraction about his archetypes which robbed them of the magic that belongs to major religious symbols. Great religious teachings were aridly generalized into manipulations of trinities or quaternities; ornate mandalas were geometrically ab-

stracted into circles and center points. There are occasions where, in spite of himself, Jung is not far removed from the callous reductionism of the French structuralists. And, at the hands of Jungian scholars, the archetypes have frequently become a basis for very conventional research—as in the Bollingen series, where the exhaustive surveys of symbolic material, with example upon example upon example, easily become a sort of higher pedantry.

Now, as psychotherapy turns more and more to spiritual tradition to find its upaya, we begin to realize that the duality of In Here–Out There will simply not do justice to religious experience. The psychological is not "merely" psychological; the unconscious, whether personal or collective, is not the termination of a journey, but a direction in which we move—on and through, as if along a dark connecting passage. Our psychotherapy is coming to grasp the subtle precisions we find in Eastern traditions and even in many "primitive" forms of worship, where careful distinctions are drawn between visionary realities and mere hallucinations or wishful thinking. These are discriminations of experience which our single, blockish category of "the unconscious" or "the subjective" cannot comprehend. John Blofeld tells us that among the Tibetan Tantrists, the deities that are served and the cosmologies that are contemplated are fully understood by the adept to be "the play of his own consciousness"; but they are not for that reason held to be "unreal." Indeed, they gain in reality by becoming more definite and enduring than empirical objects, more gracefully in tune with a reality which is experienced as fundamentally mind-like. Moreover, provided one knows the necessary upaya, these mental entities can be shared and transmitted between master and student; they can be rediscovered over and again down through the generations in dream, meditation, and trance. They emerge from a subjectivity which is transpersonal—a philosophical status with which our culture lost touch when it forgot the root meaning of "spirit."

When, during the Enlightenment, skeptical intellectuals in Western society decided there were no gods to be found among the stars or in the seas, they first concluded that all religion was a delusion or a hoax. Then Freud, drawing upon Feuerbach, contended that while religion was indeed an illusion, it was one that arose universally from the repressive dynamics of the unconscious and was then "projected" upon the external universe. Now, as we track the myths and symbols deeper and deeper into the unconscious, a third possibility arises: that the spiritual teachings are not projected *from* the mind, but *upon* it by a transpersonal reality which is the object of the mind in the same way that physical things are the objects of the senses.

4. *Universality*. The Aquarian consensus is a planetary consensus, a "symposium of the whole." Though each of us may choose a personal path and style, we cannot help but be eclectic in our appreciation of all religious and philosophical traditions, recognizing the many means that the visionary genius of our species has invented to achieve transcendence.

In the future, all exclusive claims to truth—whether those of Krishna Consciousness, Scientology, the Jesus People, or Western science—will be doomed by their own embarrassing arrogance. Openness is the only stance that befits a planetary era. Not that this need mean homogenizing all worldviews out of existence in some vast, syncretic, and featureless stew. Too much of the enlightening power we find in every salvational tradition lies in its particularity, in its peculiar and often esoteric focus upon a symbol, a ritual, a teaching. The particularity of a tradition is exactly what requires strength of commitment from its students, the willingness to spend time and learn thoroughly. It is in the particularities of a tradition that we find its finest insights. But we have learned too much about the world's culture to believe that wisdom, magnificence of spirit, and enlightenment are the monopoly of any one tradition. Great souls and great teachings arise from all corners of the planet, often to complement and balance one another brilliantly. Indeed, the spiritual life of our species has always been a network of confluent revelations appropriated and reinterpreted along subterranean lines. Search any tradition deeply enough, and it becomes a palimpsest of borrowed and reworked knowledge.

But universality is more than simple eclecticism; it must also be *convergence*. We begin by granting value to all spiritual vehicles—every rite and doctrine, ritual and teaching; we take in everything. But we finish by searching out what is humanly common and constant: the essence that has frequently been covered over by a thousand ethnic accretions. My contention here has been that an original spiritual impulse—an old gnosis—may be the source of all later religious and philosophical teachings. The image that guides me is that of the theme and its variations. Eclecticism sympathetically gathers together all the variations; universality hunts for the theme underlying them. But the theme —I am convinced—exists only as immediate, visionary insight; it cannot be directly expressed, but only translated into still another variation. And of all variations, the most valuable are those that possess the special luminosity of being transcendent symbols: declarations, rites, myths, images that uniquely serve as gateways to experience. The cross and resurrection, the jewel in the lotus, the dizzy ecstasy of the dervish, the American Indian rite of the sacred pipe, the rhapsodic prophecy, the Taoist landscape, the dance of Shiva, the sacred art and spiritual epics of the world . . . these are the great revelations; they seem to descend upon us with a power beyond human origin. They work from the source and wrest the imagination back to the source. We sense in them more vividly than any place else the heat of the original fire. What universality demands is that we search all traditions and teachings down to their symbolic level, asking what experience of life, nature, or the sacred underlies these words, these images, these rituals, these theological discriminations, until at last we reach the bridge of transcendence.

Once again, we return to the bedrock and touchstone of experience. Having discovered the universal there, we can discriminate among its million variations, enjoying them without allowing ourselves to be trapped among them.

In the past, especially in Christianity, Judaism, and Islam, the ideal of universality too frequently bogged down at an invidious and aggressive level. Yes, the idea of spiritual unity was there—but as a high-handed claim to exclusive revelation asserted belligerently by crusaders, inquisitors, or missionaries against the infidel, whose dignity had to be denigrated. Now we see that the only safeguard against such ethnocentric impudence is to search for the universal *in experience,* to treat it as a joyous, shared discovery instead of a smug, parochial privilege. And this caution applies as much to the Western world's secular ideologies as to its religions, for these too have become bellicose, intolerant causes—secular crusades directed against all spiritual tradition. Potentiality and experience (disciplined by upaya) are the elements of the Aquarian consensus that permit an honest universality. We assume transcendent potentiality as an essential human characteristic everywhere manifest, and we expect it to be unfolded by a certain quality of experience, not by the iteration of doctrines and dogmas. What we must watch for is the evidence of that experience in people's tone and conduct, in their wisdom and personal depth, in the power of their testimony. The essence of universality is to make oneself at all times the respectful student of those who are more highly evolved, wherever they may appear. One's personal spiritual commitment is not a wall that blocks them out, but a ground on which one stands as their student.

Exclusive religions—such as Christianity—may be hard-pressed to achieve a universality which is more than grudging tolerance. But that is precisely because they have lost touch with the experiential basis of their teachings and so have lost themselves in a fog of doctrine. Even the Neo-orthodox Protestants, who assert the irreducible importance of "faith," seem unable to reflect upon faith as, ultimately, some manner of validating experience that is conceivably no different as an experience than that which the Zen master and the primitive worshiper find in their commitment. Fortunately, as the mystic and esoteric veins of their tradition come back into prominence, Christians are apt to return to the visionary energies that undergird conviction, and will find, at that level, that they share a common ground with pagans and primitives, heathens and infidels, who are, after all, their fellow pilgrims in the quest.

5. *Wholeness.* I have suggested that psychopathology will be recognized in the new therapies as a wounding discrepancy between potentiality and actuality: between what we are and what we sense it is our destiny to become. Behind that primary division of the self lies a multiplicity of psychic dichotomies: passion/reason, ego/id, emotion/intellect, body/spirit. Each of these divisions represents a violation of potentiality; each is a refusal to be our whole self, an attempt to live life on reduced power. Therefore, each must, like a toxic lesion of the personality, be healed.

Currently, the effort to achieve wholeness in our society most notably takes the form of overcompensation for past repressions of sex, open emotion, aggressive self-assertion, ritual ecstasy. So we have seen, especially among the countercultural young, much emphasis on raw feeling, enthusiasm, unrefined sensation, psychedelic tripping. People search eagerly for the chance to "let it all hang out" and to voice "the primal scream." Accordingly, clear thought, articulation, and personal discipline fall into disrepute as body language, grunts, groans, and growls come to be thought of (temporarily) as the *real* us. Surely those who go this route must in time find their way back to the appropriate use of discursive reason and the well-wrought word as also illuminating powers.

But wholeness, like universality, is not simply a matter of inclusiveness. There must be an overall shape to the healthy psyche; and that shape must be an expression of what Abraham Maslow has called "hierarchical integration." It is not enough to include everything; everything must also be in its place—and its place is determined by hierarchy. The trick is to assert hierarchy *without* producing repression. And that can be achieved only where the heirarchy is dictated by nature—and so can be validated by open experience.

We have already spoken of the proper relationship in culture between the sacred triangle (myth, magic, mystery) and the profane triangle (history, technology, reason). What the ideally blended triangles represent as a balanced cultural whole is an expression of the traditional psychological hierarchy spirit/soul/body. In this threefold harmony of the personality, the task of soul and body ("soul" here including both the ego and the id functions of Freud's psychology) is to serve the transcendent needs of the spirit. In return for which, they enjoy the illumination that only spirit can provide. This integrative ideal can be discerned in the classic Indian yogas, which make place for the range of human talents: raja yoga—the yoga of visionary meditation; jnana yoga—the yoga of intellectual discrimination; bhakti yoga—the yoga of devotion, including aesthetic expression; karma yoga—the yoga of sacrificial service to others; hatha yoga—the yoga of physical energy. Many yogas, yet all are meant to be informed by a transcendent intention and to be guided by its needs. Intellect, emotion, art, ethics, organic experience have all been given room for expression. If we can imagine the yogas being practiced by one person, instead of as a number of specialties, we would see the entire conduct of life—from the immediate action of breathing through to the discharge of one's social responsibility—assuming a religious significance. No aspect of existence would need to be slighted as unworthy, fallen, or profane.

My argument has been that, at bottom, all aspects of culture—science and technology as much as art and religion, the routines of daily life as much as high ritual occasions—are elaborations of an original transcendent impulse bequeathed to us by the Few. So if we look deeply enough into any cultural exercise, we should be able to recapture its visionary core, thereby giving it its proper dignity in the hierarchy. All that wholeness must ever reject is the claim of some lesser portion of human nature to be, in itself, the whole—as when scientized rationality seeks to reduce the visionary energies to some social or physiological function and then to invent substitutes.

6. *Organicism.* The peculiar cultural style of modern Western society has been its "wordly asceticism"—to use Max Weber's phrase. As secular and historical as our society has been in its orientation, it has nevertheless (a scattering of scandalously hedonistic artists excepted) taken little joy in the body or in physical nature; it has treated both with neglect, if not hostility. Western society has been

taught by its Judeo-Christian heritage to take the world *seriously* as the realm of God's providence; yet, paradoxically, it has not been permitted to find sensual pleasure, let alone redeeming epiphanies, at the organic level. "Nature," in our tradition, is the antithesis of "grace," and that discriminatory dualism has fallen on nothing more heavily than the body. As a result, we have poured a furious energy into our worldly tasks, enough to reshape the face of the planet; but the dominant spirit of the enterprise has been ascetic in the extreme, demanding a merciless repression of the body. Thus, in all the time since the end of the Greco-Roman era, Western society has been uniquely without any native tradition of body yoga, physical culture, or athletics. So, too, with the exception of Hermeticism (always a shadowy and marginal school of thought), we have been without a continuing tradition of nature mysticism. It is precisely this ascetic style which has been—as both Weber and Freud recognized—the secret of our capitalist-industrial achievement. Only in our society have men been so willing to undertake a grueling, joyless, self-denying exploitation of themselves, others, and nature at large; to turn their whole being into a grim, anal-aggressive machinery for the accumulation of profit and power vastly beyond anything they could enjoy. Indeed, the more abstractly huge the quantities at stake, the greater the eagerness to sacrifice health, leisure, and pleasure for their possession.

We have had to learn from modern psychiatry how dear a price we pay in sanity for this despotic abuse of the body. Just as we have had to wait for new Human Potentials therapies, such as Sensory Awareness and Structural Integration, to show us what capacity the organism possesses to please, enlighten, cure, and elevate. Finally, from the yogas and physical culture of the East, we have learned how the body may be used as an instrument of meditation and transcendence. The lesson has become a permanent part of the Aquarian consensus and has given a strikingly organic quality to its tastes: a pervasive sense of unashamed physicality and whole-body eroticism. We are coming once again—as in the worldview of the Hidden Wisdom—to experience the body as a microcosm of the universe, a contemplative object whose rhythms and fibers resonate to that sensitive cosmic network which the Hermetic philosophers called the *anima mundi*. In the coming generation, one of the great challenges confronting Western medicine and psychiatry will be assimilating the esoteric anatomy we have learned from the spiritual traditions to our own scientific knowledge. For we are not only the physiologist's meat machine; we are also the mystic's celestial body of nadis and cakras. The body can be known introspectively as well as objectively, rhapsodically as well as analytically. And known from within, it is a magical instrument, a visionary resource.

Once study and respect the body in this way, and all nature around us is bound to be touched by the same enchantment, because the body is only nature nearest home. As we know the physical organism, so we will know the whole physical universe. If we discover a vital and ethereal intelligence in the workings of the body, we will soon enough discover it in the world at large. The result will be an ecological wisdom grounded in a sort of new paganism, which once again views the universe as a habitation of divine beings and intelligences whom we can know as person confronting person.

In the Western world, there has traditionally been no surer way to gain admittance to Bedlam than to go about making conversation with the dead and stupid things of the Earth—unless, of course, one was, like St. Francis, a licensed fool of God. But within the Aquarian consensus, nothing will be regarded as a clearer sign of insanity than our now "normal" inability to hear the language of Brother Sun, Sister Moon.

7. *Illumination of the Commonplace.* Among the most urgent dichotomies which the quest for wholeness will have to unify is the plaguing division between sacred and profane. And this we see happening as our society draws with an eager spontaneity upon the Zen-Taoist teaching of the illuminated commonplace: the realization that gems of enlightenment may be found in the most seemingly nondescript corners of the world. "O divine labor! Drawing water, cutting wood." So runs a Zen aphorism, reminding us that right mindfulness can find its miracles anywhere. Or, in the words of the Laotian Buddhist master Achaan Chaa, "It is all Dharma. When you do your chores, try to be mindful. If you are emptying a spittoon or cleaning a toilet, don't feel you are doing it as a favor for anyone else. There is Dharma in emptying spittoons."

All along the Aquarian frontier, we find people cultivating a ready eye for the visionary possibilities of all they encounter and do: the rhythms of the body, the chores of daily life, the practical arts and crafts, the vulgarities of popular culture. It is not fortuitous that countercultural America has brought us a renaissance of handicraft and folk art. The interest is no mere hobby or pastime; rather for many who take up these humble, long-neglected crafts, the work and the materials are filled with a rough magic. They focus contemplative attention and put the craftsman's hands joyously in touch with raw, organic stuff: wood and wool, leather and hemp, shell and bone. The finished product is frequently left . . . unfinished, for there is a special delight in preserving the native and homely qualities of things, the natural grains and Earthy textures. Or best of all, in making one's art out of scraps and cast-off objects: the treasures of the garbage dump. There were alchemists, after all, who taught that the philosopher's stone was really the dust underfoot.

This is a new aesthetic in our culture—and more than an aesthetic. Its ultimate purpose is not simply to find beauty in what was once considered lowly, but to uncover something of the great sacred Way of things in what has normally been dismissed as profane—to find it there, even there, perhaps especially there. So, too, much of the permissive eroticism of the day, insofar as it is unforced and flowing, is not simple hedonism. Rather, there is about it a certain unpretentious enchantment with organic display and pleasure, a celebration of animal naturalness as part of the Tao. As Allen Ginsberg once wrote, nicely capturing the spirit of it, "the asshole is also holy."

When one speaks of "transcendence," the idea that seems frequently to come to mind is that of a great world-denying leap into an elevated dimension of reality above and beyond the Earth. But that is a corruption of the concept as it has been taught since time out of mind. At the very origins of human worship, the sacred was experienced in, or *through*, the common stuff and substance of life: in the animals of the hunt, in the fecundity of women, in the rebirth of Spring, in the dance of the fire, in the majestic presence of the Sun. It was seen everywhere and in all things, as an object of wonder or of terror. Transcendence, in such an experience of the world, means *per-ception:* seeing through to the empowered presence of things, the ground and glory of their naked being. All that need be transcended is the illusion of ordinariness, which is only an opacity of the jaded and corrupted senses.

While Judaism, Christianity, and Islam were later to condemn this magical experience of the world as "idolatry," the knowledge of the illuminated commonplace survived in the

West in various underground traditions, for indeed it is an ancient and irrepressible truth. In Hermetic lore, it took the form of alchemical rites that worked to resurrect life eternal out of dead and vile matter. In the Kabbalistic schools, it became the doctrine of the "vestiges," the sparks of primordial divinity that were scattered and hidden throughout creation. Blake, a deep student of the Hidden Wisdom, helped pass the experience on to the Romantics; the teaching appears throughout his work, wherever he reminds us that "everything that lives is holy."

> And every space smaller than a globule of man's blood opens
> Into eternity, of which the vegetable Earth is but a shadow.

Now this ancient style of experience returns to us as a new celebration of the common, the humble, the unadorned. It is a teaching which a material simplicity of life (as well as a democratic politics) clearly demands. For in this transcendent awareness, we discover the *new wealth* of a new culture: a wealth that cannot be stolen, exploited, or counterfeited. Here is indeed the secret of the philosopher's stone which could turn dross to gold: it is, after all, a trick of the mind, a magic of the clairvoyant eye.

It should be obvious that the more widely this new wealth comes to be honored, the less we will covet conventional affluence: the merchandise, the material plenty, the cash hoards and capital which nations now kill and connive to gain. The result need not be an austere and ascetic regimen of life (though we seem to be gaining a new respect for the disciplines that have always been part of spiritual culture) but only a taste for plenitude—the sufficiency that liberates from need and greed.

Perhaps the saints have always been in touch with such truths; but when have so many ordinary mortals, having enjoyed and outgrown the charms of material abundance, been so open to the great teaching? When has it been more imperative for all the world's people to learn the limits of economic growth, the treacheries of affluence? We begin to see, in this marriage of the sacred and the profane, the seeds of a new and timely politics.

8. *Satsang.* Among those who discover the new wealth —or perhaps even so much as a glimmer of it—the need for community is bound to assert itself sooner or later as a necessity of life: the need to form sheltered and psychically nourishing environments where work, friendship, family, and the transcendent impulse interpenetrate. Especially as they come to know the richness of the spiritual traditions, they will want more than an academic once-over-lightly or a periodic dose of eclectic yoga at the nearby Growth Center. They will want continuing, concentrated instruction in the company of sympathetic minds. They will want the constant association of gurus and companions who can give their lives a significant shape. In the Buddhist tradition, this is called satsang, the true fellowship of souls striving for mutual enlightenment; we might call it eupsychian community. Possibly, out of that need, something like a new, free-wheeling, Western monasticism will be improvised among us: voluntary societies of seekers that will take the place of the rapidly decaying institutions of family, neighborhood, occupational peer group, profession, trade union. Already there are those in our society who begin to owe more heart-felt allegiance to their ashram or their therapeutic group than to any other social form. Among my own acquaintances, there are some who speak of "group" with all the warmth and commitment I would otherwise have associated with the blood family.

These therapeutic companionships are bound to be of many kinds, serving many traditions and sensibilities. The ways of personal salvation have always been multitudinous, even within religious establishments that have aspired to order and orthodoxy. The medieval Church of Rome became "Catholic" precisely because it sought to house so many ethnic and national congregations, along with the many varieties of monasticism. How much more varied are the choices apt to be in the cultural chaos of our own day, when the quest for salvation is a free adventure of the individual with all history and the world to choose from?

It would take a chapter to list only those communitarian experiments now on the American scene that attach themselves to some religious tradition or occult school. In my own area of the country alone, the San Francisco Bay Area (admittedly a supersaturated environment, but one whose innovations carry across the land), we have had, in recent years, voluntary communities connected with Tibetan Nyingmapa Buddhism, Sufism, Jesus Freak Christianity, Krishna Consciousness, Zen, Ananda Marga Yoga, Judaic Chabad, the One World Family Crusade, Ananda Meditation, the Maharaj Ji's Divine Light Mission, the mystical Judaism of Yeshivah, Integral Yoga, the teachings of Meher Baba, Gurdjieff and Ouspensky, Sun Myung Moon's Unified Family, Yogi Bhajan's Healthy-Happy-Holy Organization, the fire worship of Kailas Shugendo, the Cultural Integration Fellowship.[2] Americans now comprise one of the world's largest Zen monastic populations (in California); and in Colorado, Vermont, and California, lamas in exile are directing burgeoning Tibetan Buddhist communities which may yet see young Americans become the unlikely heirs of an ancient religious tradition now being expunged from its homeland by the legions of Chairman Mao.

The efforts I mention are notably non-Western, with few

2. One can get some idea of the variety from two recent handbooks: *The Spiritual Community Guide* (Box 1080, San Rafael, California, 94902) and the *Year One Catalog* (New York: Harper & Row, 1972).

exceptions. But our own culture may yet contribute to the options. I have for some time felt certain that a few of the Catholic religious orders—especially the Trappists or a renewed mendicant Franciscanism—might become prodigiously successful in attracting American young people into a form of lay communalism, *if* the orders could get out from under their cautious bureaucracies, recapture the spirit of their mystical origins, risk some brave social involvements, and, in general, get with the fact that a vast religious awakening is happening all about them. Perhaps the thoroughly unofficial Catholic Workers will yet gain the support their Church owes them and become a major communitarian movement. Are there no more Thomas Mertons to be found in the Church? No more Lanzo del Vastas, Peter Marins, or Dorothy Days? Is there no one in the mainstream of the Christian clergy who troubles to ask—and I do not mean only rhetorically—why it is that so many Americans turn to the exotic cults and alien traditions we find among us?

Beyond the specifically religious groups I mention, there are living or working associations which evolve out of several contemporary psychotherapies: Arica, Psychosynthesis, Transactional Analysis, Radical Therapy, Erhard Seminar Training, and any number of eclectic combinations. No doubt Synanon should also be mentioned here. Beginning about twenty years ago with the purpose of rehabilitating drug addicts, the organization has at last elaborated itself into a large, permanent, land- and property-owning community in California which has begun to attract the membership of a significant number of nonaddicts ("squares") to its regimen, lore, and lifestyle. And, significantly, at the center of the community stands the "Stew Temple," the arena of the Synanon game now at last elaborated into a therapeutic religion. From such origins, too, spiritual community can spring.

No one could know enough about all these efforts to make an intelligent evaluation of more than a few. I do not mention them here to recommend them broadcast, but only to mark the trend. Some may be shallow or opportunistic and doomed to an early demise. Doubtless all count among their following a number of transient samplers who are (understandably) trying out this and that before making any serious commitments. But the overall picture should be clear enough. The quest for satsang is on, and the options are beginning to multiply as teachers and leaders appear to meet the need. Because that need is urgent and real, I suspect it will finally winnow much of the wheat from the chaff among these experiments, leaving, for the most part, those that are robust and proven therapeutic vehicles.

What will they all have in common as expressions of the higher sanity? Three features that will outweigh the many differences and will make these many experiments a single and significant force for cultural renewal.

Each will represent a healthy collective subtraction of allegiance from the demonic ethos of power, privilege, and profit that dominates our world.

Each will assert the primacy of comradeship and participative community as the essential reality of social life.

Each will aspire to awaken the transcendent energies of its members.

Whether for any given individual this constellation of features pivots on a commitment to Sufism or Franciscan mysticism, Psychosynthesis or Zen, is of secondary importance. That is a matter of personal style which does not alter the fact that here is a common culture exerting a single, marked political force upon our world: the force of *creative disintegration*, the making of healthy communities from the human remnants of moribund institutions.

I have couched this description of the next reality in the

future tense—as a prediction. But of course it is more in the nature of a wish. Where we deal in such imponderables, what can one really do but map out the possibilities and speak for or against them? What reason have I to hope that the higher sanity stands any better chance in our time against the insolent evils that have for so long tormented the Earth? Only that the stakes are so very much higher, and the choices we make infinitely more decisive than ever before. Because urban-industrial society acts on so global a scale, an air of finality broods over all we do. The bomb is the symbol of that apocalyptic condition: the showdown, the crossroads to which we can never return to take the other path. We choose for all the future, once and for all. If nothing else, desperation—wisely doctored—may give us the courage to do the unprecedented.

Urban-industrial culture now rings us in on all sides like a man-made Himalayan range of steel and concrete. But within the interstices of that imprisoning fastness, millions of spiritually desolate lives are at work like so many rivulets of water eroding those imperious heights, weakening them at critical points, hollowing them out, opening seams for larger cascades. Ours is not the first society to experience this creative disintegration. In its later days, especially as its style of command grew more rigid and life-denying, the Roman empire was secretly fissioning into a congeries of fraternal *collegia*, mystery cults, protofeudal manorial economies, Christian congregations and monastic orders, all of which dealt more feelingly in the life-and-death realities of daily survival than Caesar's decadent institutions. The wars and deadly reason of state went on; political conniving and civil unrest continued to be the business as usual of the world—at least, at the well-lit level of reality with which our mass media would today be geared to function. But an intimate revolution was nevertheless under way in the catacombs, in the hinterlands, and in the secret chambers of the human soul. So too, in our own time, we may yet see our fossilizing industrial civilization become honeycombed with such spiritual subversions: therapeutic associations, visionary communities, in the cities and on the land, trading off affluence for autonomy, moving further and further into the Aquarian frontier, beyond the reach of the technocracy's imperatives, becoming the new society within the shell of the old.

Between the Demon and the Star
　　marking time
　　　expecting the worst
　　　　watching for a miracle

The tide is out
the inlet almost motionless beneath
　　　　　　　the early fog

gray light　gray water

I walk beside the torpid morning surf
　　wrapped in rain
　　　cloud-shrouded
The horns of unseen ships
bellow across from Georgia Strait.
Now and again, shadows of seabirds
　　　　　　sail wing-spread
through the veiling mist
come and vanish
like sudden prophecies

                                        I have it on good
                                                  authority
                                        that west of this
                                        cloud-bound coast
                                        there are islands
                                        beyond the islands

My thoughts—as always—broil with

reports of the daily terror.
I pay too much attention to the
                    latest news,
let the seasons of eternity slip
                by and by

Yes, I know
the fierce convictions that tear
                the Earth.
are impassioned ignorance, bubbles
                of angry ego.
But the suffering of people is real
the victims are real, one by one by one

I worry for my daughter, little girl
                            lost
in the forests of war
I cannot shake my mind free . . .   But let me tell you
                                         what I've heard
                                         about the islands

cannot meditate my way
around the cries of children
my head is always busy with . . .   If I could only
                                                  tell you
                                         of the islands

I wish I could feed my mind
to these wheeling gulls, let them
                              carry it
across the gray waves, find light
                      and peace
on the far side of this cloud   O, islands
                                                   islands

let my attention grow still and
                          perfect
in the still and perfect light

a point of concentrated flame

a star

This morning
from where I stand

I can hardly see the islands
nearest shore

Vancouver, 1975

Index of Names